# Information Security and Cryptography
## Texts and Monographs

*Series Editors*
David Basin
Ueli Maurer

*Advisory Board*
Martín Abadi
Ross Anderson
Michael Backes
Ronald Cramer
Virgil D. Gligor
Oded Goldreich
Joshua D. Guttman
Arjen K. Lenstra
John C. Mitchell
Tatsuaki Okamoto
Kenny Paterson
Bart Preneel

For further volumes:
http://www.springer.com/series/4752

Rainer Böhme

# Advanced Statistical Steganalysis

 Springer

Dr. Rainer Böhme
Technische Universität Dresden
Fakultät Informatik
Institut für Systemarchitektur
01062 Dresden
Germany
rainer.boehme@tu-dresden.de

*Series Editors*

Professor Dr. Ueli Maurer
Professor Dr. David Basin
ETH Zürich
Switzerland
maurer@inf.ethz.ch
basin@inf.ethz.ch

ISSN 1619-7100
ISBN 978-3-642-14312-0          e-ISBN 978-3-642-14313-7
DOI 10.1007/978-3-642-14313-7
Springer Heidelberg Dordrecht London New York

Library of Congress Control Number: 2010933722

ACM Computing Classification (1998): E.3, C.3, H.5.1, I.4

*Cover design*: KuenkelLopka GmbH

Printed on acid-free paper

Springer is a part of Springer Science+Business Media (www.springer.com)

*To Siegfried*

# Foreword

Steganography and steganalysis, the hiding and detection of a covert payload within an innocent cover object, started to receive attention from the computer science, engineering, and mathematics communities in the 1990s. At first the problems were not clearly defined, but proper statistical foundations were proposed and mathematical rigour has gradually entered the literature.

After an explosion of interest around the turn of the century, both fields have developed apace, with numerous research papers appearing in proceedings of conferences such as the Information Hiding Workshops and International Workshops on Digital Watermarking, published by Springer, and the ACM Multimedia Security Workshop, SPIE Electronic Imaging, and the new IEEE Workshop on Information Forensics and Security, published by their respective learned societies. There are also new journals dedicated to information hiding topics. But such a wide field is difficult to browse for a reader who is not involved in active research, and this book aims to collect some of the key advances under a common theme. It is suitable for a knowledgable scientist who is not necessarily an expert in information hiding, and it begins with an ab initio exposition of the aims and techniques of steganography and steganalysis. The reader can hope to gain a general understanding of the field as well as of some specific steganalysis techniques.

One particular difficulty for the lay reader of published research is the unfortunate tendency of papers to contradict one other. For example, one reads of a "perfectly secure stego system" in one paper, which shortly afterwards is shown to be reliably unmasked. How can such apparently contradictory statements be reconciled? The answer lies in the assumptions which have been made about the innocent cover objects in which the hidden data is smuggled: commonly, the assumptions are not even stated at all, let alone precisely, and mismatches of assumptions lead to contradictory conclusions. Indeed, most researchers now agree that the covers are the most important part of the steganography and steganalysis battle, and that understanding covers properly is most important to advancing the field. A chapter of this book is devoted to clarifying what we understand by a "cover model," and

separating the different approaches. The reader may then peruse the literature more critically, understanding the limitations of proposed methods.

The book also contains four examples of theoretical advances in statistical steganalysis, including completely some novel detection methods as well
as advancing standing methods using new techniques. Each example is validated by simulations and experiments with large sets of genuine covers. The
advances include a detector for so-called "model-based steganography" using
first-order statistics, remarkable because model-based steganography was introduced with the claim of perfect security against first-order steganalysis;
a new methodology to study the effects of heteogeneity in cover images by
an analysis of payload estimation error; a new application of the so-called
"WS" method to more sensitive detection in never-compressed and JPEG
pre-compressed covers; and a method to identify the encoder, and hence improve steganalysis, in MP3 audio files.

This material is presented using consistent notation throughout, including
a presentation of some of my work on "structural steganalysis." Some of the
included technical advances were inspired by collaborations between myself
and the author, who shares my preference for a mathematical explanation of
why steganalysis works over ad hoc proposals justified only empirically. It is
rewarding to see these ideas come to fruition in an extended text.

Oxford, October 2009                                              *Andrew D. Ker*

# Acknowledgements

The author is indebted to the following colleagues who supported him in the preparation of this book. Andrew Ker has been a very valuable discussions partner, and collaboration with him yielded substantial parts of Chapters 5 and 6. Further, he gave insightful comments on drafts of this manuscript and generously shared with me ideas, cover images and data generated with his code and computing resources. So it was appropriate for him to contribute the Foreword to this book. Andreas Pfitzmann offered me the opportunity and encouraged me to continue my research and write the dissertation [17] on which this book is based. I also enjoyed discussions with Jessica Fridrich, an attentive reader and critic of my research papers, on various occasions. In particular, the idea to further decompose the within-image error into two components is based on her remarks. Matthias Kirchner commented on the entire draft. It also benefited from further comments on specific parts by Dagmar Schönfeld and Antje Winkler. Andreas Westfeld drew my attention to steganography and steganalysis, which he presented in his lecture as an exciting field of research. Motivated by some of his preliminary results, I started the work on MP3 classification, for which he could eventually raise funds so that the research behind Chapter 7 could be completed in a joint project. Last but not least, I am grateful to all my colleagues in the Privacy and Data Security research group in Dresden for numerous hints and for creating a stimulating atmosphere for research. The amount of experimental evidence in this book would not have been manageable without the computing resources of TU Dresden's ZIH and the Oxford University Computing Laboratory. The responsibility for all remaining errors and omissions is solely with the author.

Berkeley, October 2009                                     *Rainer Böhme*

# Contents

## Part II Specific Advances in Steganalysis

# Chapter 1
# Introduction

## 1.1 Steganography and Steganalysis as Empirical Sciences

Steganography is the ancient art and young science of hidden communication. A broad definition of the subject includes all endeavours to communicate in such a way that the existence of the message cannot be detected. Unlike cryptography, which merely ensures the *confidentiality* of the message content, steganography adds another layer of secrecy by keeping confidential even the fact that secret communication takes place. The corresponding protection goal is called *undetectability* [192].

It follows from the definition of steganography as security technique that its quality cannot be assessed without coming to conclusions on how difficult it is to detect the presence of a hidden message, i.e., to break steganography. As a result, progress in steganography is closely tied to advances in *steganalysis*, the science of detecting hidden information. The two aspects can hardly be studied separately.

With the development of digital communication, *digital steganography* has emerged as a science that deals with inconspicuously embedding digital messages in other digital or digitised data, so-called *covers*. It is a subfield of *information hiding*, which comprises digital steganography, covert channels, digital watermarking and the subset of privacy-enhancing technologies built on the data avoidance principle, together with their respective detection and counter-technologies [193].

A common approach for steganographic methods to be used in open communication systems is to conceal secret messages in covers which are transmitted through the communication system as normal messages. Undetectability is reached if steganographic traffic cannot be distinguished from normal 'plausible' traffic. In other words, the goal of steganographic systems is to generate output that is equal to *something outside the system*. This implies that we cannot succeed in designing secure steganography without studying the outside of the system! So we conclude that steganography and steganalysis, according to their common definition and in the predominant setting, should be considered as genuinely empirical disciplines, unlike cryptology and related fields of information security. This may sound inconvenient from a purely mathematical or theoretical standpoint, but it makes steganography and steganalysis particularly interesting engineering problems.

Nevertheless, it is not wise to abandon the realm of theory entirely by approaching steganography and steganalysis as pure inductive disciplines. So, in the course of this book, we refer to information theory as well as statistical theory and combine them with epistemology to lay theoretical foundations on which the engineering problems can be formulated. Such foundations also allow us to deduct when and under which assumptions we can expect solutions in terms of security guarantees, and when we cannot. Such insight helps to reappraise discoveries in practical steganalysis—prior art and our own.

## 1.2 Objective and Approach

This work sets out to provide an account of recent advances in statistical steganalysis with a special emphasis on the role of empirical covers in steganography. Let us briefly specify two keywords of this statement: cover and statistical steganalysis.

Covers play a key role in the interpretation of steganography and steganalysis as empirical sciences. They link a steganographic system to its outside world. Most steganographic algorithms take as input covers which are digital signals that represent parts of reality, such as image or audio data. Compared to cryptography, the possibility to choose a cover could be seen as an extra degree of freedom that promises advantages to the steganographer over the steganalyst. Conversely, having the cover generation outside of the steganographer's trusted domain may also add a new source of uncertainty, and hence a security risk. As covers are the 'unknowns', the least controllable part in steganography and steganalysis, the best one can do to study their influence in practical applications is to build and validate *models* of covers. This creates a link to statistical signal analysis.

Advanced steganalysis almost always draws on statistical methods to decide whether an observed signal, which superficially resembles a typical cover, actually contains a secret message or not. This methodological approach suggests that covers should be regarded as realisations of random variables. In particular, the distribution assumptions of these random variables deserve attention to avoid undue generalisations, but are frequently neglected. While early statistical steganalysis was tested on single—possibly selected—example covers, it soon became common practice to test both steganographic and steganalytic methods on sets in the range of dozens of covers and report summary statistics. This already was a modest improvement upon single tests on the omnipresent 'Lena' image,[1] but it is still insufficient for two reasons. First, the test images were often obtained from a single homogeneous source without careful control of possible preprocessing. Second, to calculate summary measures, standard statistical tools have often been applied without deeper reflection on their underlying assumptions and conditions (such as homogeneity). Both practices risk spurious results, because—this is a guiding principle supported with evidence in this book—most empirical covers are not homogeneous.

As a consequence, this book presents ways to appropriately deal with the heterogeneity in statistical models on various levels of analysis. To do this, we develop a consistent theoretical framework which puts models of covers and their imperfections as decisive elements in the race between steganography and steganalysis. Although the idea to formulate steganalysis problems in terms of statistical models is not entirely novel (e.g., [206]), most model refinements in the literature have led to specialisation, so the applicability of results was limited to ever more specific cases. By contrast, we introduce the notion of conditional cover models to deal with heterogeneity in the model-based framework. This *can* be seen as a kind of meta-model that governs a set of more specific models. Mixture cover models, based on the statistical tool of mixture distributions, are proposed as an elegant way to think of cover-specific steganalysis. With regard to the system architecture for practical steganalysis, this concept paves the way to break up monolithic detectors into more modular ones.

From an academic point of view, the framework provides an umbrella to four specific instances of concrete steganalysis problems, on which we study heterogeneity between covers, between sources of covers, and between preprocessing histories of covers. In three of these instances we achieve improvements of steganalysis performance, i.e., we make better decisions when detecting secret messages. In one case, a new detection method could be found against a steganographic method that has previously been believed secure.

As with every empirical research, new insight leads to new open questions. So it is apparent that this work cannot cover the outlined field in its entirety. Nor can we control all combinations of conditions in every experiment. However, whenever appropriate, the most promising directions for further investigations in our opinion are pointed out. A more detailed exposition of the individual results is given in the next section, together with an introduction to the structure of this book.

---

[1] A standard test image used in the image processing literature, http://www.petitcolas.net/fabien/watermarking/image_database/lena.jpg.

## 1.3 Outline

This book is structured in three parts. Part I covers general aspects, including a survey of the state of the art, and our theoretical framework. Part II documents four self-contained specific advances in steganalysis research. Each of these results demonstrates how selected aspects of our theoretical conclusions can yield measurable performance gains in relevant practical situations. Part III combines general insights and specific findings and concludes with an outlook to possible future developments.

The first part, entitled 'Background and Advances in Theory', is divided into two chapters. Chapter 2 contains a review of the state of the art: almost two decades of research on steganography and relevant related aspects of information security, signal processing and compression. It thereby builds the essential background to understand the research presented in the following chapters. The presentation largely draws on established definitions and terms, harmonised throughout the book, so that readers who are familiar with the field can safely skip Chapter 2 without impairing the comprehensibility of the remainder.

Among the two options for an internal structure of a didactic review of related work, top-down and bottom-up, we have followed the latter. Although top-down is often perceived as a more elegant solution, and certainly appropriate for many topics, we have decided against it, because it would require us to start with theoretical secure systems and descend gradually to practical solutions. This would have been appropriate for cryptography, where theoretical security is achievable. But this is not the case in steganography (some pathological cases aside). Despite bottom-up, we try to present the methods and issues involved in practical steganography and steganalysis in as modular and general a way as possible: Sections 2.1 to 2.3 introduce the basic communication model, our notation and conventions, a classification of design goals as well as the associated metrics. Sections 2.4 and 2.5 distinguish general design principles and classes of adversaries, respectively. Section 2.6 deals with covers and various options to represent them in data structures. It recalls the relevant details of file formats of the covers studied in this book. Further, low-level embedding operations are central to Section 2.7, before higher-level coding and protocol aspects are discussed in Section 2.8. The remainder of the chapter is devoted to steganalysis. Section 2.9 presents specific detection techniques, and the penultimate Section 2.10 explains in detail four quantitative detectors for LSB replacement steganography and their variants. This class of detectors is further studied and extended in Chapters 5 and 6 in the second part of this work. Section 2.11 summarises and concludes the survey. Overall, Chapter 2 offers a fairly comprehensive overview of the area with one single exception: the literature on information-theoretic approaches has intentionally been excluded. The reason for this is partly because this literature is less accessible, due to formalisms, and of limited practical relevance. But the main reason is that it fits better into the next chapter.

Chapter 3 builds the theoretical underpinning for this work and contributes advances in theory. We recombine existing ideas with novel ones to build a coherent framework which

1. clarifies the link to epistemology for what we call empirical covers (Sect. 3.1),
2. sharpens the notion of statistical models for cover signals and emphasises their role in the iterative chase for supremacy between steganography and steganalysis (Sect. 3.2),
3. suggests a novel way of thinking of heterogeneous cover sources by means of conditional cover models (Sect. 3.3),
4. proposes a system to structure existing approaches to secure steganography by cover assumptions and adversary assumptions (Sect. 3.4), and
5. is applicable to practical problems (Sect. 3.5): all specific findings in the second part are instances of 'model problems' studied in Chapter 3.

Therefore, at the end of Chapter 3, equipped with all relevant background of modern steganographic and steganalytic techniques, we are in a position to apply the theoretical framework to practical problems. This guides us to Part II, entitled 'Specific Advances in Steganalysis', which has four chapters.

Commensurate with the topic of this book, we have taken a cross-sectional view on an increasingly diverse research field. While it was possible to make fairly general statements on the abstract level of analysis in Part I, in particular Chapter 3, it is impossible to maintain the breadth when deepening the level of analysis in Part II. Instead, we drill down into four applied steganalysis research questions, which can be framed as instances of the aspects of the theory, but at the same time constitute self-contained studies of relevance, and with interesting results. Chapters 4 to 7 are based on a selection of our published research [15, 18, 22, 23, 25, 26, 133], though rewritten in large parts and adapted to the new terminology and context. In some cases, the results have been rearranged and are presented in more detail than in the original research papers. Also, the conclusions have been revised to be more balanced against the backdrop of recent findings, with more distance, or simply from a better overview of the area (this applies particularly to the parts first published in 2004). So, our hope is that even the specific chapters are valuable to readers who have already read our publications.

Chapter 4 revisits the model-based approach to steganography [206] in the light of our theory. MB1, the first *steganographic algorithm* designed along these principles, is one of the few that employ an *explicit cover model*, in this case for JPEG images. We can show that this model is too much an idealisation of real covers and thus, when used in MB1, produces unnaturally homogeneous histograms of quantised DCT coefficients. A statistical test for model compliance applied to all tails of AC DCT coefficient histograms allows us to distinguish clean covers from stego objects by counting the number of outliers. The presence of outliers in natural covers is a sign of heterogeneity in the signal distributions. Results from experiments with a modest set of test images show that this new detector achieves good performance under various conditions. This is particularly remarkable, as our detector only uses first-order statistics, ignoring all spatial relations between samples, and thus refutes the claim that MB1 is secure against steganalysis based on first-order statistics.

Chapter 5 studies the character and source of heterogeneity in cover images by an analysis of estimation errors of quantitative detectors. Detectors of this kind are based on a set of cover assumptions, which jointly form a cover model in our terminology. Deviations of true covers from these assumptions lead to errors in the estimation of the (presumed) secret message length. We propose a statistical methodology, which takes into account the specific heavy-tailed distribution of estimation errors between covers, to calculate more meaningful performance metrics than previously used in the literature. In addition, the methodology can be extended to regress explanatory variables on the size of the estimation error. This allows us to relate steganalytic performance to macroscopic image properties and, when such properties can be measured in practice, to reduce the error due to heterogeneity when making a steganalysis decision. The usefulness of the methodology is demonstrated with results from five different quantitative detectors on a large set of images.

Chapter 6 presents improvements of the steganalysis method based on a weighted stego image [73], one of the few *steganalysis methods* that employ an *explicit cover model*. On the one hand, our improvements target the cover model for never-compressed images, which results in noteworthy performance gains (Sect. 6.1). On the other hand, we demonstrate the usefulness of conditional cover models by designing a cover model for images which had been pre-compressed to JPEG before having been used as covers in steganographic communication. (Such cases are in fact relevant, as many acquisition devices, foremost digital cameras, store as JPEG by default.) While pre-compressed images used to be a weak spot of the methodology, we can show that the specialised cover model boosts performance by one order of magnitude, thus leaving behind the best structural detectors, which were known to be more robust to JPEG pre-compression (Sect. 6.2). Again, experimental results on large image sets and several robustness checks back our evidence.

Chapter 7 demonstrates once more how conditional cover models can improve the reliability of steganalysis. We pick MP3-compressed audio files as covers, for which the encoder implementation is a relevant source of heterogeneity: different implementations of the same encoding standard produce distinct output streams for the same uncompressed input. The steganalysis method to detect `MP3Stego` [184, 234] already contains a sufficiently good (implicit) cover model, but its compliance with real covers strongly depends on the encoder. So, instead of developing a conditional cover model, the challenge in this chapter is to estimate the condition, i.e., whether the existing model is applicable or not, in a specific steganalysis case. We present a method to distinguish between MP3 encoders based on ten statistical features that can be extracted from the compressed streams. The method employs a Bayesian machine learning classifier to determine the most likely encoder of a set of 20 encoders. Experimental evidence shows that the method is reliable enough to substantially reduce the error rates in steganalysis.

The third part, 'Synthesis', summarises the results of both Parts I and II, and points to open research questions on a more abstract level than possible from the perspective of the individual specific studies. We also take the opportunity to draw parallels to similar questions in related fields.

# Part I
# Background and Advances in Theory

# Chapter 2
# Principles of Modern Steganography and Steganalysis

The first work on digital steganography was published in 1983 by cryptographer Gustavus Simmons [217], who formulated the problem of steganographic communication in an illustrative example that is now known as the *prisoners' problem*[1]. Two prisoners want to cook up an escape plan together. They may communicate with each other, but all their communication is monitored by a warden. As soon as the warden gets to know about an escape plan, or any kind of scrambled communication in which he suspects one, he would put them into solitary confinement. Therefore, the inmates must find some way of hiding their secret messages in inconspicuous cover text.

## 2.1 Digital Steganography and Steganalysis

Although the general model for steganography is defined for arbitrary communication channels, only those where the cover media consist of multimedia objects, such as image, video or audio files, are of practical relevance.[2] This is so for three reasons: first, the cover object must be large compared to the size of the secret message. Even the best-known embedding methods do not allow us to embed more than 1% of the cover size securely (cf. [87, 91] in conjunction with Table A.2 in Appendix A). Second, indeterminacy[3] in the cover is necessary to achieve steganographic security. Large objects without indeterminacy, e.g., the mathematical constant $\pi$ at very high precision, are unsuitable covers since the warden would be able to verify their regular

---

[1] The prisoners' problem should not be confused with the better-known prisoners' dilemma, a fundamental concept in game theory.

[2] Artificial channels and 'exotic' covers are briefly discussed in Sects. 2.6.1 and 2.6.5, respectively.

[3] Unless otherwise stated, *indeterminacy* is used with respect to the uninvolved observer (warden) throughout this book. The output of indeterministic functions may be deterministic for those who know a (secret) internal state.

structure and discover traces of embedding. Third, transmitting data that contains indeterminacy must be plausible. Image and audio files are so vital nowadays in communication environments that sending such data is inconspicuous.

As in modern cryptography, it is common to assume that Kerckhoffs' principle [135] is obeyed in digital steganography. The principle states that the steganographic algorithms to embed the secret message into and extract it from the cover should be public. Security is achieved solely through secret keys shared by the communication partners (in Simmons' anecdote: agreed upon before being locked up). However, the right interpretation of this principle for the case of steganography is not always easy, as the steganographer may have additional degrees of freedom [129]. For example, the selection of a cover has no direct counterpart in standard cryptographic systems.

## 2.1.1 Steganographic System

Figure 2.1 shows the baseline scenario for digital steganography following the terminology laid down in [193]. It depicts two parties, sender and recipient, both steganographers, who communicate covertly over the public channel. The sender executes function $\mathsf{Embed} : \mathcal{M} \times \mathcal{X}^* \times \mathcal{K} \to \mathcal{X}^*$ that requires as inputs the secret message $m \in \mathcal{M}$, a plausible cover $x^{(0)} \in \mathcal{X}^*$, and the secret key $k \in \mathcal{K}$. $\mathcal{M}$ is the set of all possible messages, $\mathcal{X}^*$ is the set of covers transmittable over the public channel and $\mathcal{K}$ is the key space. $\mathsf{Embed}$ outputs a stego object $x^{(m)} \in \mathcal{X}^*$ which is indistinguishable from (but most likely not identical to) the cover. The stego object is transmitted to the recipient who runs $\mathsf{Extract} : \mathcal{X}^* \times \mathcal{K} \to \mathcal{M}$, using the secret key $k$, to retrieve the secret message $m$. Note that the recipient does not need to know the original cover to extract the message. The relevant difference between covert and encrypted communication is that for covert communication it is hard or impossible to infer the *mere existence* of the secret message from the observation of the stego object without knowledge of the secret key.

The combination of embedding and extraction function for a particular type of cover, more formally the quintuple $(\mathcal{X}^*, \mathcal{M}, \mathcal{K}, \mathsf{Embed}, \mathsf{Extract})$, is called *steganographic system*, in short, *stego system*.[4]

---

[4] This definition differs from the one given in [253]: Zhang and Li model it as a sextuple with separate domains for covers and stego objects. We do not follow this definition because the domain of the stego objects is implicitly fixed for given sets of covers, messages and keys, and two transformation functions. Also, we deliberately exclude distribution assumptions for covers from our system definition.

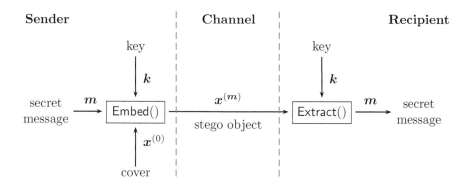

Fig. 2.1: Block diagram of baseline steganographic system

## 2.1.2 Steganalysis

The security of a steganographic system is defined by its strength to defeat detection. The effort to detect the presence of steganography is called *steganalysis*. The steganalyst (i.e., the warden in Simmons' anecdote) is assumed to control the transmission channel and watch out for suspicious material [114]. A steganalysis method is considered as successful, and the respective steganographic system as 'broken', if the steganalyst's decision problem can be solved with higher probability than random guessing [33].

Note that we have not yet made any assumptions on the computational complexity of the algorithms behind the functions of the steganographers, Embed and Extract, and the steganalyst's function Detect : $\mathcal{X}^* \to$ {cover, stego}. It is not uncommon that the steganalyst's problem can theoretically be solved with high probability; however, finding the solution requires vast resources. Without going into formal details, the implicit assumption for the above statements is that for an operable steganographic system, embedding and extraction are computationally easy whereas reliable detection requires considerably more resources.

## 2.1.3 Relevance in Social and Academic Contexts

The historic roots of steganography date back to the ancient world; the first books on the subject were published in the 17th century. Therefore, the art is believed to be older than cryptography. We do not repeat the phylogenesis of covert communication and refer to Kahn [115], Petitcolas et al. [185]

or, more comprehensively, Kipper [139, Chapter 3], who have collected numerous examples of covert communication in the pre-digital age. Advances in modern digital steganography are relevant for academic, engineering, national security and social reasons. For society at large, the existence of secure steganography is a strong argument for the opponents of crypto regulation, a debate that has been fought in Germany in the 1990s and that reappears on the agendas of various jurisdictions from time to time [63, 142, 143]. Moreover, steganographic mechanisms can be used in distributed peer-to-peer networks that allow their users to safely evade Internet censorship imposed by authoritarian states. But steganography is also a 'dual use' technique: it has applications in defence, more precisely in covert field communication and for hidden channels in cyber-warfare tools. So, supposedly intelligence agencies are primarily interested in steganalysis. Steganography in civilian engineering applications can help add new functionality to legacy protocols while maintaining compatibility (the security aspect is subordinated in this case) [167]. Some steganographic techniques are also applicable in digital rights management systems to protect intellectual property rights of media data. However, this is mainly the domain of digital watermarking [42], which is related to but adequately distinct from pure steganography to fall beyond the scope of this book. Both areas are usually subsumed under the term 'information hiding' [185].[5] Progress in steganography is beneficial from a broader academic perspective because it is closely connected to an ever better understanding of the stochastic processes behind cover data, i.e., digital representations of natural images and sound. Refined models, for whatever purpose, can serve as building blocks for better compression and recognition algorithms. Steganography is interdisciplinary and touches fields of computer security, particularly cryptography, signal processing, coding theory, and machine learning (pattern matching). Steganography is also closely conected (both methodologically but also by an overlapping academic community) to the emerging field of multimedia forensics. This branch develops [177] and challenges [98, 140] methods to detect forgeries in digital media.

## 2.2 Conventions

Throughout this book, we use the following notation. Capital letters are reserved for random variables $X$ defined over the domain $\mathcal{X}$. Sets and multisets are denoted by calligraphic letters $\mathcal{X}$, or by double-lined capitals for special sets $\mathbb{R}$, $\mathbb{Q}$, $\mathbb{Z}$, etc. Scalars and realisations of random variables are printed in lower case, $x$. Vectors of $n$ random variables are printed in boldface (e.g.,

---

[5] Information hiding as a subfield of information security should not be confused with information hiding as a principle in software engineering, where some authors use this term to describe techniques such as abstract data types, object orientation, and components. The idea is that lower-level data structures are hidden from higher-level interfaces [181].

$\boldsymbol{X} = (X_1, X_2, \ldots, X_n)$ takes its values from elements of the product set $\mathcal{X}^n$). Vectors and matrices, possibly realisations of higher-dimensional random variables, are denoted by lower-case letters printed in boldface, $\boldsymbol{x}$. Their elements are annotated with a subscript index, $x_i$ for vectors and $x_{i,j}$ for matrices. Subscripts to boldface letters let us distinguish between realisations of a random vector; for instance, $\boldsymbol{m}_1$ and $\boldsymbol{m}_2$ are two different secret messages. Functions are denoted by sequences of characters printed in sans serif font, preceded by a capital letter, for example, $\mathsf{F}(x)$ or $\mathsf{Embed}(\boldsymbol{m}, \boldsymbol{x}^{(0)}, \boldsymbol{k})$.

No rule without exception: we write $\boldsymbol{k}$ for the key, but reuse scalar $k$ as an index variable without connection to any element of a vector of key symbols. Likewise, $N$ is used as alternative constant for dimensions and sample sizes, not as a random variable. $\boldsymbol{I}$ is the identity matrix (a square matrix with 1s on the main diagonal and 0s elsewhere), not a random vector. Also $\mathcal{O}$ has a double meaning: as a set in sample pair analysis (SPA, Sect. 2.10.2), and elsewhere as the complexity-theoretic Landau symbol $\mathcal{O}(n)$ with denotation 'asymptotically bounded from above'.

We use the following conventions for special functions and operators:

- **Set theory**   $\mathfrak{P}$ is the power set operator and $|\mathcal{X}|$ denotes the cardinality of set $\mathcal{X}$.
- **Matrix algebra**   The inverse of matrix $\boldsymbol{x}$ is $\boldsymbol{x}^{-1}$; its transposition is $\boldsymbol{x}^\mathsf{T}$. The notation $\boldsymbol{1}_{i \times j}$ defines a matrix of 1s with dimension $i$ (rows) and $j$ (columns). Operator $\otimes$ stands for the Kronecker matrix product or the outer vector product, depending on its arguments. Operator $\odot$ denotes element-wise multiplication of arrays with equal dimensions.
- **Information theory**   $\mathsf{H}(X)$ is the Shannon entropy of a discrete random variable or empirical distribution (i.e., a histogram). $\mathsf{D}_{\mathsf{KL}}(X, Y)$ is the relative entropy (Kullback–Leibler divergence, KLD [146]) between two discrete random variables or empirical distributions, with the special case $\mathsf{D}_{\mathsf{bin}}(u, v)$ as the binary relative entropy of two distributions with parameters $(u, 1 - u)$ and $(1 - v, v)$. $\mathsf{D}_{\mathsf{H}}(\boldsymbol{x}, \boldsymbol{y})$ is the Hamming distance between two discrete sequences of equal length.
- **Probability calculus**   $\mathsf{Prob}(x)$ denotes the probability of event $x$, and $\mathsf{Prob}(x|y)$ is the probability of $x$ conditionally on $y$. Operator $\mathsf{E}(X)$ stands for the expected value of its argument $X$. $X \sim \mathcal{N}(\mu, \sigma)$ means that random variable $X$ is drawn from a Gaussian distribution with mean $\mu$ and standard deviation $\sigma$. Analogously, we write $\mathcal{N}(\boldsymbol{\mu}, \boldsymbol{\Sigma})$ for the multivariate case with covariance matrix $\boldsymbol{\Sigma}$. When convenient, we also use probability spaces $(\Omega, \mathcal{P})$ on the right-hand side of operator '$\sim$', using the simplified notation $(\Omega, \mathcal{P}) = (\Omega, \mathfrak{P}(\Omega), \mathcal{P})$ since the set of events is implicit for countable sample spaces. We write the uniform distribution over the interval $[a, b]$ as $\mathcal{U}_a^b$ in the continuous case and as $\ddot{\mathcal{U}}_a^b$ in the discrete case (i.e., all integers $i : a \leq i \leq b$ are equally probable). Further, $\mathcal{B}(n, \pi)$ stands for a binomial distribution as the sum of $n$ Bernoulli trials over $\{0, 1\}$ with probability to draw a 1 equal to $\pi$. Unless otherwise stated,

the hat annotation $\hat{x}$ refers to an estimate of a true parameter $x$ that is only observable indirectly through realisations of random variables.

We further define a special notation for embedded content and write $\boldsymbol{x}^{(0)}$ for cover objects and $\boldsymbol{x}^{(1)}$ for stego objects. If the length of the embedded message is relevant, then the superscript may contain a scalar parameter in brackets, $\boldsymbol{x}^{(p)}$, with $0 \leq p \leq 1$, measuring the secret message length as a fraction of the total capacity of $\boldsymbol{x}$. Consistent with this convention, we write $\boldsymbol{x}^{(i)}$ if it is uncertain, but not irrelevant whether $\boldsymbol{x}$ represents a cover or a stego object. In this case we specify $i$ further in the context. If we wish to distinguish the content of multiple embedded messages, then we write $\boldsymbol{x}^{(m_1)}$ and $\boldsymbol{x}^{(m_2)}$ for stego objects with embedded messages $\boldsymbol{m}_1$ and $\boldsymbol{m}_2$, respectively. The same notation can also be applied to elements $x_i$ of $\boldsymbol{x}$: $x_i^{(0)}$ is the $i$th symbol of the plain cover and $x_i^{(1)}$ denotes that the $i$th symbol contains a steganographic semantic. This means that this symbol is used to convey the secret message and can be interpreted by Extract. In fact, $x_i^{(0)} = x_i^{(1)}$ if the steganographic meaning of the cover symbol already matches the respective part of the message. Note that there is not necessarily a one-to-one relation between message symbols and cover symbols carrying secret message information $x_i^{(1)}$, as groups of cover symbols can be interpreted jointly in certain stego systems (cf. Sect. 2.8.2).

Without loss of generality, we make the following assumptions in this book:

- The secret message $\boldsymbol{m} \in \mathcal{M} = \{0,1\}^*$ is a vector of bits with maximum entropy. (The Kleene closure operator $*$ is here defined under the vector concatenation operation.) We assume that symbols from arbitrary discrete sources can be converted to such a vector using appropriate source coding. The length of the secret message is measured in bits and denoted as $|\boldsymbol{m}| \geq 0$ (as the absolute value interpretation of the $|x|$ operator can be ruled out for the message vector). All possible messages of a fixed length appear with equal probability. In practice, this can be ensured by encrypting the message before embedding.
- Cover and stego objects $\boldsymbol{x} = (x_1, \ldots, x_n)$ are treated as column vectors of integers, thus disregarding any 2D array structure of greyscale images, or colour plane information for colour images. So, we implicitly assume a homomorphic mapping between samples in their spatial location and their position in vector $\boldsymbol{x}$. Whenever the spatial relation of samples plays a role, we define specific mapping functions, e.g., Right : $\mathbb{Z}^+ \to \mathbb{Z}^+$ between the indices of, say, a pixel $x_i$ and its right neighbour $x_j$, with $j = \mathsf{Right}(i)$. To simplify the notation, we ignore boundary conditions when they are irrelevant.

## 2.3 Design Goals and Metrics

Steganographic systems can be measured by three basic criteria: capacity, security, and robustness. The three dimensions are not independent, but should rather be considered as competing goals, which can be balanced when designing a system. Although there is a wide consensus on the same basic criteria, the metrics by which they are measured are not unanimously defined. Therefore, in the following, each dimension is discussed together with its most commonly used metrics.

### 2.3.1 Capacity

Capacity is defined as the maximum length of a secret message. It can be specified in *absolute* terms (bits) for a given cover, or as *relative* to the number of bits required to store the resulting stego object. The capacity depends on the embedding function, and may also depend on properties of the cover $x^{(0)}$. For example, least-significant-bit (LSB) replacement with one bit per pixel in an uncompressed eight-bit greyscale image achieves a net capacity of 12.5%, or slightly less if one takes into account that each image is stored with header information which is not available for embedding. Some authors would report this as 1 bpp (*bits per pixel*), where the information about the actual bit depths of each pixel has to be known from the context. Note that not all messages are maximum length, so bits per pixel is also used as a measure of capacity usage or *embedding rate*. In this work, we prefer the latter term and define a metric $p$ (for 'proportion') for the length of the secret message relative to the maximum secret message length of a cover. Embedding rate $p$ has no unit and is defined in the range $0 \leq p \leq 1$. Hence, for an embedding function which embeds one bit per cover symbol,

$$p = \frac{|m|}{n} \quad \text{for covers} \quad x^{(0)} \in \mathcal{X}^n. \tag{2.1}$$

However, finding meaningful measures for capacity and embedding rate is not always as easy as here. Some stego systems embed into compressed cover data, in which the achievable compression rate may vary due to embedding. In such cases it is very difficult to agree on the best denominator for the capacity calculation, because the size of the cover (e.g., in bytes, or in pixels for images) is not a good measure of the amount of information in a cover. Therefore, specific capacity measures for particular compression formats of cover data are needed. For example, F5, a steganographic algorithm for JPEG-compressed images, embeds by *decreasing* the file size almost monotonically with the amount of embedded bits [233]. Although counterintuitive at first sight, this works by reducing the image quality of the lossy compressed image

Table 2.1: Result states and error probabilities of a binary detector

|                     | Reality | |
| --- | --- | --- |
| **Detector output** | plain cover | stego object |
| plain cover | correct rejection<br>$1 - \alpha$ | miss<br>$\beta$ |
| stego object | false positive<br>$\alpha$ | correct detection<br>$1 - \beta$ |

further below the level of distortion that would occur without steganographic content. As a result, bpc (*bits per nonzero DCT coefficient*) has been proposed as a capacity metric in JPEG images.

It is intuitively clear, often demonstrated (e.g., in [15]), and theoretically studied[6] that longer secret messages ceteris paribus require more embedding changes and thus are statistically better detectable than smaller ones. Hence, capacity and embedding rate are related to security, the criterion to be discussed next.

## 2.3.2 Steganographic Security

The purpose of steganographic communication is to hide the mere existence of a secret message. Therefore, unlike cryptography, the security of a steganographic system is judged by the impossibility of detecting rather than by the difficulty of reading the message content. However, steganography builds on cryptographic principles for removing recognisable structure from message content, and to control information flows by the distribution of keys.

The steganalysis problem is essentially a decision problem (does a given object contain a secret message or not?), so decision-theoretic metrics qualify as measures of steganographic security and, by definition, equally as measures of steganalytic performance. In steganalysis, the decision maker is prone to two types of errors, for which the probabilities of occurrence are defined as follows (see also Table 2.1):

- The probability that the steganalyst fails to detect a stego object is called *missing probability* and is denoted by $\beta$.

---

[6] Capacity results can be found in [166] and [38] for specific memoryless channels, in Sect. 3 of [253] and [41] for stego systems defined on general artificial channels, and in [134] and [58] for stego systems with empirical covers. Theoretical studies of the trade-off between capacity and robustness are common (see, for example, [54, 172]), so it is surprising that the link between capacity and security (i.e., detectability) is less intensively studied.

- The probability that the steganalyst misclassifies a plain cover as a stego object is called *false positive probability* and denoted by $\alpha$.

Further, $1 - \beta$ is referred to as *detection probability*. In the context of experimental observations of detector output, the term 'probability' is replaced by 'rate' to signal the relation to frequencies counted in a finite sample. In general, the higher the error probabilities, the better the security of a stego system (i.e., the worse the decisions a steganalyst makes).

Almost all systematic steganalysis methods do not directly come to a binary conclusion (cover or stego), but base their binary output on an internal state that is measured at a higher precision, for example, on a continuous scale. A decision threshold $\tau$ is used to quantise the internal state to a binary output. By adjusting $\tau$, the error rates $\alpha$ and $\beta$ can be traded off. A common way to visualise the characteristic relation between the two error rates when $\tau$ varies is the so-called *receiver operating characteristics* (ROC) curve. A typical ROC curve is depicted in Fig. 2.2 (a). It allows comparisons of the security of alternative stego systems for a fixed detector, or conversely, comparisons of detector performance for a fixed stego system. Theoretical ROC curves are always concave,[7] and a curve on the 45° line would signal perfect security. This means a detector performs no better than random guessing.

One problem of ROC curves is that they do not summarise steganographic security in a single figure. Even worse, the shape of ROC curves can be skewed so that the respective curves of two competing methods intersect (see Fig. 2.2 (b)). In this case it is particularly hard to compare different methods objectively.

As a remedy, many metrics derived from the ROC curve have been proposed to express steganographic security (or steganalysis performance) on a continuous scale, most prominently,

- the *detector reliability* as area under the curve (AUC), minus the triangle below the 45° line, scaled to the interval $[0, 1]$ (a measure of *insecurity*: values of 1 imply perfect detectability) [68],
- the false positive rate at 50% detection rate (denoted by $\text{FP}_{50}$),
- the *equal error rate* $\text{EER} = \alpha \quad \Leftrightarrow \quad \alpha = \beta$,
- the *total minimal decision error* $\text{TMDE} = \min_\tau \dfrac{\alpha + \beta}{2}$ [87], and
- the minimum of a cost- or utility-weighted sum of $\alpha$ and $\beta$ whenever dependable weights are known for a particular application (for example, false positives are generally believed to be more costly in surveillance scenarios).

If one agrees to use one (and only one) of these metrics as the 'gold standard', then steganographic systems (or detectors) can be ranked according to its value, but statistical inference from finite samples remains tricky. A sort of inference test can be accomplished with critical values obtained from

---

[7] Estimated ROC curves from a finite sample of observations may deviate from this property unless a probabilistic quantiser is assumed to make the binary decision.

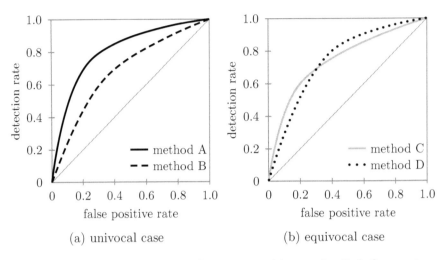

Fig. 2.2: ROC curve as measure of steganographic security. Left figure: stego system A is less secure than stego system B, because for any fixed false positive rate, the detection rate for A is higher than for B (in fact, both methods are insecure). Right figure: the relative (in)security of stego systems C and D depends on the steganalyst's decision threshold.

bootstrapping extensive simulation data, as demonstrated for a theoretical detector response in [235].

Among the list of ROC-based scalar metrics, there is no unique best option. Each metric suffers from specific weaknesses; for instance, AUC aggregates over practically irrelevant intervals of $\tau$, EER and $FP_{50}$ reflect the error rates for a single arbitrary $\tau$, and the cost-based approach requires application-specific information.

As a remedy, recent research has tried to link theoretically founded metrics of statistical distinguishability, such as the Kullback–Leibler divergence between distributions of covers and stego objects, with practical detectors. This promises more consistent and sample-size-independent metrics of the amount of evidence (for the presence of a secret message) accumulated *per* stego object [127]. However, current proposals to approximate lower bounds (i.e., guaranteed insecurity) for typical stego detectors require thousands of measurements of the detector's internal state. So, more rapidly converging approximations from the machine learning community have been considered recently [188], but it is too early to tell if these metrics will become standard in the research community.

If the internal state is not available, a simple method to combine both error rates with an information-theoretic measure is the *binary relative entropy* of two binary distributions with parameters $(\alpha, 1 - \alpha)$ and $(1 - \beta, \beta)$ [34]:

$$D_{\text{bin}}(\alpha, \beta) = \alpha \log_2 \frac{\alpha}{1 - \beta} + (1 - \alpha) \log_2 \frac{1 - \alpha}{\beta}. \tag{2.2}$$

A value of $D_{\text{bin}}(\alpha, \beta) = 0$ indicates perfect security (against a specific decision rule, i.e., detector) and larger positive values imply better detectability. This metric has been proposed in the context of information-theoretic bounds for steganographic security. Thus, it is most useful to compare relatively secure systems (or weak detectors), but unfortunately it does not allow us to identify perfect separation ($\alpha = \beta = 0$). $D_{\text{bin}}(\alpha, \beta)$ converges to infinity as $\alpha, \beta \to 0$.

Finally and largely independently, *human perceptibility* of steganographic modifications in the cover media can also be subsumed to the security dimension, as demonstrated by the class of *visual attacks* [114, 238] against simple image steganography. However, compared to modern statistical methods, visual approaches are less reliable, depend on particular image characteristics, and cannot be fully automated. Note that in the area of watermarking, it is common to use the term *transparency* to describe visual imperceptibility of embedding changes. There, visual artefacts are not considered as a security threat, because the *existence* of hidden information is not a secret. The notion of security in watermarking is rather linked to the difficulty of *removing* a mark from the media object. This property is referred to as *robustness* in steganography and it has the same meaning in both steganographic and watermarking systems, but it is definitely more vital for the latter.

## 2.3.3 Robustness

The term robustness means the difficulty of removing hidden information from a stego object. While removal of secret data might not be a problem as serious as its detection, robustness is a desirable property when the communication channel is distorted by random errors (channel noise) or by systematic interference with the aim to prevent the use of steganography (see Sect. 2.5 below). Typical metrics for the robustness of steganographic algorithms are expressed in distortion classes, such as additive noise or geometric transformation. Within each class, the amount of distortion can be further specified with specific (e.g., parameters of the noise source) or generic (e.g., peak signal-to-noise ratio, PSNR) distortion measures. It must be noted that robustness has not received much attention so far in steganography research. We briefly mention it here for the sake of completeness. The few existing publications on this topic are either quite superficial, or extremely specific [236]. Nevertheless, robust steganography is a relevant building block for the construction of secure and effective censorship-resistant technologies [145].

### 2.3.4 Further Metrics

Some authors define additional metrics, such as *secrecy*, as the difficulty of extracting the message content [253]. We consider this beyond the scope of steganographic systems as the problem can be reduced to a confidentiality metric of the cryptographic system employed to encrypt a message prior to embedding (see [12] for a survey of such metrics). The computational *embedding complexity* and the *success rate*, i.e., the probability that a given message can be embedded in a particular cover at a given level of security and robustness, become relevant for advanced embedding functions that impose constraints on the permissible embedding distortion (see Sect. 2.8.2). Analogously, one can define the *detection complexity* as the computational effort required to achieve a given combination of error rates $(\alpha, \beta)$, although even a computationally unbounded steganalyst in general cannot reduce error rates arbitrarily for a finite number of observations. We are not aware of focused literature on detection complexity for practical steganalysis.

## 2.4 Paradigms for the Design of Steganographic Systems

The literature distinguishes between two alternative approaches to construct steganographic systems, which are henceforth referred to as *paradigms*.

### 2.4.1 Paradigm I: Modify with Caution

According to this paradigm, function Embed of a stego system takes as input cover data provided by the user who acts as sender, and embeds the message by modifying the cover. Following a general belief that fewer and smaller changes are less detectable (i.e., are more secure) than more and larger changes, those algorithms are designed to carefully preserve as many characteristics of the cover as possible.

Such distortion minimisation is a good heuristic in the absence of a more detailed cover model, but is not always optimal. To build a simple counterexample, consider as cover a stereo audio signal in a frequency domain representation. A hypothetical embedding function could attempt to shift the phase information of the frequency components, knowing that phase shifts are not audible to human perception and difficult to verify by a steganalyst who is unaware of the exact positioning of the microphones and sound sources in the recording environment. Embedding a secret message by shifting $k$ phase coefficients in both channels randomly is obviously less secure than shifting $2k$ coefficients in both channels symmetrically, although the embedding distortion (measured in the number of cover symbols changed) is doubled. This is so because humans can hear phase differences between two mixing sources, and a steganalyst could evaluate asymmetries between the two channels, which are atypical for natural audio signals.

Some practical algorithms have taken up this point and deliberately modify more parts of the cover in order to restore some statistical properties that are known to be analysed in steganalytic techniques (for example, OutGuess [198] or statistical restoration steganography [219, 220]). However, so far none of the actively preserving algorithms has successfully defeated targeted detectors that search for particular traces of active preservations (i.e., evaluate other statistics than the preserved ones). Some algorithms even turned out to be less secure than simpler embedding functions that do not use complicated preservation techniques (see [24, 76, 187, 215]). The crux is that it is difficult to change all symbols in a high-dimensional cover consistently, because the entirety of dependencies is unknown for empirical covers and cannot be inferred from a single realisation (cf. Sect. 3.1.3).

## 2.4.2 Paradigm II: Cover Generation

This paradigm is of a rather theoretical nature: its key idea is to replace the cover as input to the embedding function with one that is computer-generated by the embedding function. Since the cover is created entirely in the sender's trusted domain, the generation algorithm can be modified such that the secret message is already formed at the generation stage. This circumvents the problem of unknown interdependencies because the exact cover model is implicitly defined in the cover generating algorithm (see Fig. 2.3 and cf. artificial channels, Sect. 2.6.1).

The main shortcoming of this approach is the difficulty of conceiving plausible cover data that can be generated with (indeterministic) algorithms. Note that the fact that covers are computer-generated must be plausible in the communication context.[8] This might be true for a few mathematicians or artists who exchange colourful fractal images at high definition,[9] but is less so if supporters of the opposition in authoritarian states discover their passion for mathematics. Another possible idea to build a stego system following this paradigm is a renderer for photo-realistic still images or videos that contain indeterministic effects, such as fog or particle motion, which could be modulated by the secret message. The result would still be recognisable as computer-generated art (which may be plausible in some contexts), but its

---

[8] If the sender pretended that the covers are representations of reality, then one would face the same dilemma as in the first paradigm: the steganalyst could exploit imperfections of the generating algorithm in modelling the reality.

[9] Mandelsteg is a tool that seems to follow this paradigm, but it turns out that the fractal generation is not dependent on the secret message. ftp://idea.sec.dsi.unimi.it/pub/security/crypt/code/

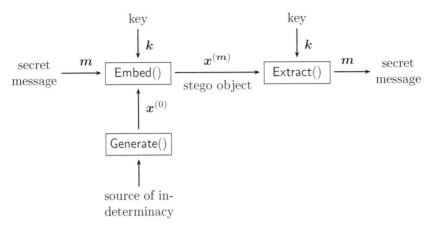

Fig. 2.3: Block diagram of stego system in the cover generation paradigm

statistical properties would not differ from similar art created with a random noise source to seed the indeterminism. Another case could be made for a steganographic digital synthesiser, which uses a noise source to generate drum and cymbal sounds.[10] Aside from the difficulty or high computational complexity of extracting such messages, it is obvious that the number of people dealing with such kind of media is much more limited than those sending digital photographs as e-mail attachments. So, the mere fact that uncommon data is exchanged may raise suspicion and thus thwart security. The only practical example of this paradigm we are aware of is a low-bandwidth channel in generated animation backgrounds for video conferencing applications, as recently proposed by Craver et al. [45].

A weaker form of this paradigm tries to avoid the plausibility problem without requiring consistent changes [64]. Instead of simulating a cover generation process, a plausible (ideally indeterministic, and at the least not invertible) cover *transformation* process is sought, such as downscaling or changing the colour depth of images, or, more general, lossy compression and redigitisation [65]. Figure 2.4 visualises the information flow in such a construction. We argue that stego systems simulating deterministic but not invertible transformation processes can be seen as those of paradigm I, 'Modify with Caution', with side information available exclusively to the sender. This is so because their security depends on the indeterminacy in the cover rather

---

[10] One caveat to bear in mind is that typical random number generators in creative software do not meet cryptographic standards and may in fact be predictable. Finding good pseudorandom numbers in computer-generated art may thus be an indication for the use of steganography. As a remedy, Craver et al. [45] call for 'cultural engineering' to make sending (strong) pseudorandom numbers more common.

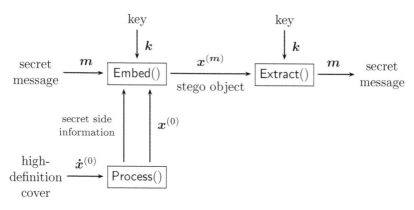

Fig. 2.4: Stego system with side information based on a lossy (or indeterministic) process: the sender obtains an information advantage over adversaries

than on artificially introduced indeterminacy (see Sect. 3.4.5 for further discussion of this distinction). Nevertheless, for the design of a stego system, the perspective of paradigm II may prove to be more practical: it is sometimes preferable for the steganographer to know precisely what the steganalyst most likely *will not know*, rather than to start with vague assumptions on what the steganalyst *might know*. Nevertheless, whenever the source of the cover is not fully under the sender's control, it is impossible to guarantee security properties because information leakage through channels unknown to the designer of the system cannot be ruled out.

## 2.4.3 Dominant Paradigm

The remainder of this chapter, in its function to provide the necessary background for the specific advances presented in the second part of this book, is confined to paradigm I, 'Modify with Caution'. This reflects the dominance of this paradigm in contemporary steganography and steganalysis research. Another reason for concentrating on the first paradigm is our focus on steganography and steganalysis in natural, that is empirical, covers. We argue in Sect. 2.6.1 that covers of (the narrow definition of) paradigm II constitute artificial channels, which are not empirical. Further, in the light of these arguments, we outline in Sect. 3.4.5 how the traditional distinction of paradigms in the literature can be replaced by a distinction of cover assumptions, namely (purely) empirical versus (partly) artificial cover sources.

## 2.5 Adversary Models

As in cryptography research, an adversary model is a set of assumptions defining the goals and limiting the computational power and knowledge of the steganalyst. Specifying adversary models is necessary because it is impossible to realise security goals against omnipotent adversaries. For example, if the steganalyst knows $x^{(0)}$ for a specific act of communication, a secret message is detectable with probability $\mathsf{Prob}\left(i \neq 0 | x^{(i)}\right) = 1 - 2^{-|m|}$ by comparing objects $x^{(i)}$ and $x^{(0)}$ for identity. The components of an adversary model can be structured as follows:

- **Goals** The stego system is formulated as a probabilistic game between two or more competing players [117, for example].[11] The steganalyst's goal is to win this game, as determined by a utility function, with non-negligible probability. (A function $\mathsf{F} : \mathbb{Z}^+ \rightarrow [0, 1]$ is called *negligible* if for every security parameter $\ell > 0$, for all sufficiently large $y$, $\mathsf{F}(y) < 1/y^{\ell}$.)[12]
- **Computational power** The number of operations a steganalyst can perform and the available memory are bounded by a function of the security parameter $\ell$, usually a polynomial in $\ell$.
- **Knowledge** Knowledge of the steganalyst can be modelled as information sets, which may contain realisations of (random) variables as well as random functions ('oracles'), from which probability distributions can be derived through repeated queries (sampling).

From a security point of view, it is useful to define the strongest possible, but still realistic, adversary model. Without going into too many details, it is important to distinguish between two broad categories of adversary models: *passive* and *active* warden.[13]

### 2.5.1 Passive Warden

A passive warden is a steganalyst who does not interfere with the content on the communication channel, i.e., who has read-only access (see Fig. 2.5). The steganalyst's goal is to correctly identify the existence of secret messages by running function Detect (not part of the stego system, but possibly adapted to a specific one), which returns a metric to decide if a specific $x^{(i)}$ is to be

---

[11] See Appendix E for an example game formulation (though some terminology is not introduced yet).

[12] Note that this definition does not limit the specification of goals to 'perfect' security (i.e., the stego system is broken if the detector is marginally better than random guessing). A simple construction that allows the specification of bounds to the error rates is a game in which the utility is cut down by the realisation of a random variable.

[13] We use the terms 'warden' and 'steganalyst' synonymously for steganographic adversaries. Other substitutes in the literature are 'attacker' and 'adversary'.

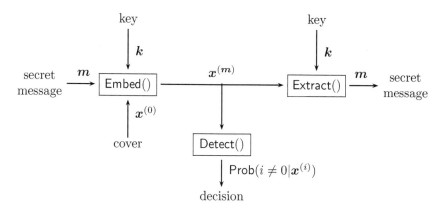

Fig. 2.5: Block diagram of steganographic system with passive warden

considered as a stego object or not. A rarely studied extension of this goal is to create evidence which allows the steganalyst to prove to a third party that steganography has been used.

Some special variants of the passive warden model are conceivable:

- Ker [123, 124] has introduced *pooled steganalysis*. In this scenario, the steganalyst inspects a set of suspect objects $\{x_1^{(i_1)}, \ldots, x_N^{(i_N)}\}$ and has to decide whether steganography is used in any of them or not at all. This scenario corresponds to a situation where a storage device, on which secret data may be hidden in anticipation of a possible confiscation, is seized. In this setting, sender and recipient may be the same person. Research questions of interest deal with the strategies to distribute secret data in a batch of $N$ covers, i.e., to find the least-detectable sequence $(i_1, \ldots, i_N)$, as well as the optimal aggregation of evidence from $N$ runs of Detect.
- Combining multiple outcomes of Detect is also relevant to *sequential steganalysis* of an infinite stream of objects $(x_1^{(i_1)}, x_2^{(i_2)}, \ldots)$, pointed out by Ker [130]. Topics for study are, again, the optimal distribution $(i_1, i_2, \ldots)$, ways to augment Detect by a memory of past observations $\mathsf{Detect}^{\mathcal{P}} : \mathfrak{P}(\mathcal{X}^*) \to \mathbb{R}$, and the timing decision about after how many observations sufficient evidence has accumulated.
- Franz and Pfitzmann [65] have studied, among other scenarios, the so-called *cover–stego-attacks*, in which the steganalyst has some knowledge $\hat{x}^{(0)}$ about the cover of a specific act of communication, but not the exact realisation $x^{(0)}$. This happens, for example, if a cover was scanned from a newspaper photograph: both sender and steganalyst possess an analogue copy, so the information advantage of the sender over the steganalyst is

merely the noise introduced in his private digitising process. Another example is embedding in MP3 files of commercially sold music.

- A more ambitious goal of a passive warden than detecting the presence of a secret message is learning its content. Fridrich et al. [84] discuss how the detector output for specific detectors can be used to identify likely stego keys.[14] This is relevant because the correct stego key cannot be found by exhaustive search if the message contains no recognisable redundancy, most likely due to prior encryption (with an independent crypto key). A two-step approach via the stego key can reduce the complexity of an exhaustive search for both stego and crypto keys from $\mathcal{O}(2^{2\ell})$ to $\mathcal{O}(2^{\ell+1})$ (assuming key sizes of $\ell$ bits each). Information-theoretic theorems on the secrecy of a message (as opposed to security $\leftrightarrow$ detectability) in a stego system can be found in [253].

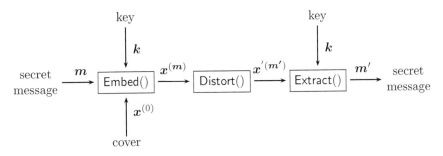

Fig. 2.6: Block diagram of steganographic system with active warden

### 2.5.2 Active Warden

In the active warden model, a steganalyst has read and write access to the communication channel. The wardens's goal is to prevent hidden communication or impede it by reducing the capacity of the hidden channel. This can be modelled by a distortion function $\mathsf{Distort} : \mathcal{X}^* \to \mathcal{X}^*$ in the communication channel (see Fig. 2.6). Note that systematic distortion with the aim to corrupt stego objects may also affect legitimate use of the communication channel adversely (e.g., by introducing visible noise or artefacts). Conversely, common transformations on legitimate channels may, as a side effect, distort

---

[14] We distinguish between 'stego' and 'crypto' keys only with regard to the secrecy of the message content: the former secures the fact that a message is present and the latter secures its content.

steganography despite not being designed with this intention (e.g., JPEG recompression or scaling on public photo communities or auction websites). Active warden models fit in the above-discussed structure for adversary models by specifying the warden's goals in a multistage game in which the options for the steganographers depend on previous moves of the warden.

Again, some variants of the active warden model are worth mentioning:

- A steganalyst, whose goal is to detect the use of steganography, could be in a position to supply the cover, or alter its value, before it is used as input to Embed by the sender. This happens, for example, when the steganalyst sells a modified digitisation device to the suspect sender, which embeds two watermarks in each output $x^{(0)}$: one is robust against changes introduced by Embed and the other is fragile [155]. The use of steganography can be detected if an observed object $x^{(i)}$ contains the robust watermark (which ensures that the tampered device has actually been used as the cover source), but not the fragile one (the indication that an embedding function as been applied on the cover). The robust watermark, which is a harder problem to realise, can be omitted if the fact that the cover is taken from the tampered device can be inferred from the context.

- A steganalyst can also actively participate as pretended communication partner in multiphase protocols, such as a covert exchange of a public stego key in public-key steganography (PKS). Consider a protocol where two communication partners perform a 'stego handshake' by first passing a public key embedded in a stego object $x_1^{(k_{\mathrm{pub}})}$ from the sender (initiator) to the recipient, who uses it to encrypt a message that is returned in a stego object $x_2^{(\mathrm{Encrypt}(m, k_{\mathrm{pub}}))}$. An active warden could act as initiator and 'challenge' a suspect recipient with a public-key stego object. The recipient can be convicted of using steganography if the reply contains an object from which a message with verifiable redundancy can be extracted using the respective private key. This is one reason why it is hard to build secure high capacity public-key steganography with reasonable cover assumptions[15] in the active warden model.

In practical applications we may face a combination of both passive and active adversaries. Ideal steganography thus should be a) secure to defeat passive steganalysis and b) robust to thwart attempts of interference with covert channels. This links the metrics discussed in Sect. 2.3 to the adversary models. The adversary model underlying the analyses in the second part of this book is the passive warden model.

---

[15] In particular, sampling cover symbols conditional on their history is inefficient. Such constructions have been studied by Ahn and Hopper [3] and an extension to adaptive active adversaries has been proposed by Backes and Cachin [8]. Both methods require a so-called 'rejection sampler'.

## 2.6 Embedding Domains

Before we drill down into the details of functions Embed and Extract in Sects. 2.7 and 2.8, respectively, let us recall the options for the domain of the cover representation $\mathcal{X}^*$. To simplify the notation, we consider covers $\mathcal{X}^n$ of finite dimension $n$.

### 2.6.1 Artificial Channels

Ahead of the discussion of empirical covers and their domains relevant to practical steganography, let us distinguish them from *artificial covers*. Artificial covers are sequences of elements $x_i$ drawn from a *theoretically defined* probability distribution over a discrete channel alphabet of the underlying communication system. There is no uncertainty about the parameters of this distribution, nor about the validity of the cover model. The symbol generating process *is* the model. In fact, covers of the (strong form of) paradigm II, 'Cover Generation', are artificial covers (cf. Sect. 2.4).

We also use the term *artificial channel* to generalise from individual cover objects to the communication system's channel, which is assumed to transmit a sequence of artificial covers. However, a common simplification is to regard artificial covers of a single symbol, so the distinction between artificial channels and artificial covers can be blurry. Another simplification is quite common in theoretical work: a channel is called *memoryless* if there are no restrictions on what symbol occurs based on the history of channel symbols, i.e., all symbols in a sequence are independent. It is evident that memoryless channels are well tractable analytically, because no dependencies have to be taken into account.

Note that memoryless channels with known symbol distributions can be efficiently compressed to full entropy random bits and vice versa.[16] Random bits, in turn, are indistinguishable from arbitrary cipher text. In an environment where direct transmission of cipher text is possible and tolerated, there is no need for steganography. Therefore we deem artificial channels not relevant covers in practical steganography. Nevertheless, they do have a raison d'être in theoretical work, and we refer to them whenever we discuss results that are only valid for artificial channels.

The distinction between empirical covers and artificial channels resembles, but is not exactly the same as, the distinction between *structured* and *unstructured* covers made by Fisk et al. [60]. A similar distinction can also be found in [188], where our notion of artificial channels is called

---

[16] In theory, this also applies to stateful (as opposed to memoryless) artificial channels with the only difference being that the compression algorithm may become less efficient.

*analytical model*, as opposed to *high-dimensional model*, which corresponds to our notion empirical covers.[17]

## 2.6.2 Spatial and Time Domains

*Empirical covers* in spatial and time domain representations consist of elements $x_i$, which are discretised samples from measurements of analogue signals that are continuos functions of location (space) or time. For example, images in the spatial domain appear as a matrix of intensity (brightness) measurements sampled at an equidistant grid. Audio signals in the time domain are vectors of subsequent measurements of pressure, sampled at equidistant points in time (sampling rate). Digital video signals combine spatial and time dimensions and can be thought of as three-dimensional arrays of intensity measurements.

Typical embedding functions for the spatial or time domain modify individual sample values. Although small changes in the sample intensities or amplitudes barely cause perceptual differences for the cover as a whole, spatial domain steganography has to deal with the difficulty that spatially or temporally related samples are not independent. Moreover, these multivariate dependencies are usually non-stationary and thus hard to describe with statistical models. As a result, changing samples in the spatial or time domain consistently (i.e., preserving the dependence structure) is not trivial.

Another problem arises from file format conventions. From an information-theoretic point of view, interdependencies between samples are seen as a redundancy, which consumes excess storage and transmission resources. Therefore, common file formats employ lossy source coding to achieve leaner representations of media data. Steganography which is not robust to lossy coding would only be possible in uncompressed or losslessly compressed file formats. Since such formats are less common, their use by steganographers may raise suspicion and hence thwart the security of the covert communication [52].

## 2.6.3 Transformed Domain

A time-discrete signal $\boldsymbol{x} = (x_1, \ldots, x_n)$ can be thought of as a point in $n$-dimensional space $\mathbb{R}^n$ with a Euclidean base. The same signal can be expressed in an infinite number of alternative representations by changing the base. As long as the new base has at least rank $n$, this transformation is invertible and no information is lost. Different domains for cover representations are defined

---

[17] We do not follow this terminology because it confounds the number of dimensions with the empirical or theoretical nature of cover generating processes. We believe that although both aspects overlap often in practice, they should be separated conceptually.

by their linear transformation matrix $\boldsymbol{a}$: $\boldsymbol{x}_{\text{trans}} = \boldsymbol{a}\,\boldsymbol{x}_{\text{spatial}}$. For large $n$, it is possible to transform disjoint sub-vectors of fixed length from $\boldsymbol{x}$ separately, e.g., in blocks of $N^2 = 8 \times 8 = 64$ pixels for standard JPEG compression.

Typical embedding functions for the transformed domain modify individual elements of the transformed domain. These elements are often called 'coefficients' to distinguish them from 'samples' in the spatial domain.[18]

Orthogonal transformations, a special case, are rotations of the $n$-dimensional coordinate system. They are linear transformations defined by orthogonal square matrices, that is, $\boldsymbol{a}\,\boldsymbol{a}^{\mathsf{T}} = \boldsymbol{I}$, where $\boldsymbol{I}$ is the identity matrix. A special property is that Euclidean distances in $\mathbb{R}^n$ space are invariant to orthogonal transformations. So, both embedding distortion and quantisation distortion resulting from lossy compression, measured as *mean square error* (MSE), are invariant to the domain in which the distortion is introduced.

Classes of orthogonal transformations can be distinguished by their ability to decorrelate elements of $\boldsymbol{x}$ if $\boldsymbol{x}$ is interpreted as a realisation of a random vector $\boldsymbol{X}$ with nonzero covariance between elements, or by their ability to concentrate the signal's energy in fewer (leading) elements of the transformed signal. The energy of a signal is defined as the square norm of the vector $e_{\boldsymbol{x}} = ||\boldsymbol{x}||$ (hence, energy is invariant to orthogonal transformations). However, both the optimal decorrelation transformation, the Mahalanobis transformation [208], as well as the optimal energy concentration transformation, the Karhunen–Loeve transformation [116, 158], also known as principal component analysis (PCA), are signal-dependent. This is impractical for embedding, as extra effort is required to ensure that the recipient can find out the exact transformation employed by the sender,[19] and not fast enough for the compression of individual signals. Therefore, good (but suboptimal) alternatives with fixed matrix $\boldsymbol{a}$ are used in practice.

The family of *discrete cosine transformations* (DCTs) is such a compromise, and thus it has a prominent place in image processing. A 1D DCT of column vector $\boldsymbol{x} = (x_1, \ldots, x_N)$ is defined as $\boldsymbol{y} = \boldsymbol{a}_{1\mathrm{D}}\,\boldsymbol{x}$, with elements of the orthogonal matrix $\boldsymbol{a}_{1\mathrm{D}}$ given as

$$a_{ij} = \sqrt{\frac{2}{N}} \cdot \cos\left(\frac{(2j-1)(i-1)\pi}{2N}\right)\left(1 + \frac{\delta_{i,1}}{2}(\sqrt{2} - 2)\right), \quad 1 \leq i, j \leq N. \tag{2.3}$$

Operator $\delta_{i,j}$ is the Kronecker delta:

$$\delta_{i,j} = \begin{cases} 1 \text{ for } i = j \\ 0 \text{ for } i \neq j. \end{cases} \tag{2.4}$$

---

[18] We use 'sample' as a more general term when the domain does not matter.

[19] Another problem is that no correlation does not imply independence, which can be shown in a simple example. Consider the random variables $X = \sin\omega$ and $Y = \cos\omega$ with $\omega \sim \mathcal{U}_0^{2\pi}$; then, $\text{cor}(X, Y) \propto \mathsf{E}(XY) = \int_0^{2\pi} \sin u \cos u \, du = 0$, but $X$ and $Y$ are dependent, for example, because $\mathsf{Prob}(x = 0 \pm \varepsilon) < \mathsf{Prob}(x = 0|y = 1) = 1/2$, $\varepsilon^2 \ll 1$. So, finding an uncorrelated embedding domain does not enable us to embed consistently with *all* possible dependencies between samples.

$(4, 4)$                                            $\mathbf{a}_{2D}$

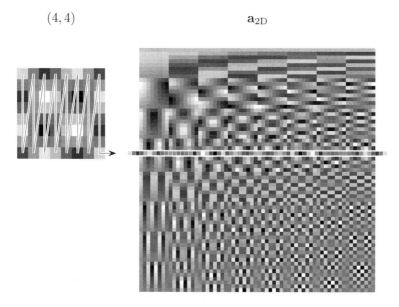

Fig. 2.7: $8 \times 8$ blockwise DCT: relation of 2D base vectors (example: subband $(4, 4)$) to row-wise representation in the transformation matrix $\boldsymbol{a}_{2D}$

Two 1D-DCT transformations can be combined to a linear-separable 2D-DCT transformation of square blocks with $N \times N$ elements. Let all $k$ blocks of a signal $\boldsymbol{x}$ be serialised in columns of matrix $\boldsymbol{x}_{\boxplus}$; then,

$$\boldsymbol{y}_{\boxplus} = \boldsymbol{a}_{2D}\, \boldsymbol{x}_{\boxplus} \quad \text{with}$$
$$\boldsymbol{a}_{2D} = \left(\mathbf{1}_{N \times 1} \otimes \boldsymbol{a}_{1D} \otimes \mathbf{1}_{1 \times N}\right) \odot \left(\mathbf{1}_{1 \times N} \otimes \boldsymbol{a}_{1D} \otimes \mathbf{1}_{N \times 1}\right). \tag{2.5}$$

Matrix $\boldsymbol{a}_{2D}$ is orthogonal and contains the $N^2$ base vectors of the transformed domain in rows. Figure 2.7 illustrates how the base vectors are represented in matrix $\boldsymbol{a}_{2D}$ and Fig. 2.8 shows the typical DCT base vectors visualised as $8 \times 8$ intensity maps to reflect the 2D character. The base vectors are arranged by increasing the horizontal and vertical spatial frequency subbands.[20] The upper-left base vector $(1, 1)$ is called the DC (direct current) component; all the others are AC (alternating current) subbands. Matrix $\boldsymbol{y}_{\boxplus}$ contains the transformed coefficients in rows, which serve as weights for the $N^2$ DCT base vectors to reconstruct the block in the inverse DCT (IDCT),

$$\boldsymbol{x}_{\boxplus} = \boldsymbol{a}_{2D}^{-1}\, \boldsymbol{y}_{\boxplus} = \boldsymbol{a}_{2D}^{\mathsf{T}}\, \boldsymbol{y}_{\boxplus}. \tag{2.6}$$

---

[20] Another common term for 'spatial frequency subband' is 'mode', e.g., in [189].

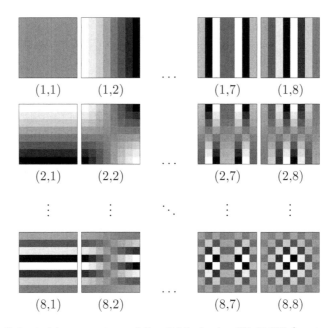

(1,1)        (1,2)        ...        (1,7)        (1,8)

(2,1)        (2,2)        ...        (2,7)        (2,8)

(8,1)        (8,2)        ...        (8,7)        (8,8)

Fig. 2.8: Selected base vectors of $8 \times 8$ blockwise 2D DCT (vectors mapped to matrices)

In both $\boldsymbol{x}_\boxplus$ and $\boldsymbol{y}_\boxplus$, each column corresponds to one block. Note that a direct implementation of this mathematically elegant single transformation matrix method would require $\mathcal{O}(N^4)$ multiplication operations per block of $N \times N$ samples. Two subsequent 1D-DCT transformations require $\mathcal{O}(2N^3)$ operations, whereas *fast DCT* (FDCT) algorithms reduce the complexity further by factorisation and use of symmetries down to $\mathcal{O}(2N^2 - N \log_2 N - 2N)$ multiplications per block [57] (though this limit is only reachable at the cost of more additions, other trade-offs are possible as well).

Other common transformations not detailed here include the *discrete Fourier transformation* (DFT), which is less commonly used because the resulting coefficients contain phase information in the imaginary component of complex numbers, and the *discrete wavelet transformation* (DWT), which differs from the DCT in the base functions and the possibility to decompose a signal hierarchically at different scales.

In contrast to DCT and DFT domains, which are constructed from orthogonal base vectors, the *matching pursuit* (MP) 'domain' results from a decomposition with a highly redundant basis. Consequently, the decomposition is not unique and heuristic algorithms or other tricks, such as side information from related colour channels (e.g., in [35]), must be used to ensure that both sender and recipient obtain the same decomposition path before and after embedding. Embedding functions operating in the MP domain, albeit barely tested with targeted detectors, are claimed to be more secure than spatial domain embedding because changes appear on a 'higher semantic level' [35, 36].

Unlike spatial domain representations in the special case of natural images, for which no general statistical model of the marginal distribution of intensity values is known, distributions of AC DCT coefficients tend to be unimodal and symmetric around 0, and their shape fits Laplace (or more generally, Student $t$ and Generalised Gaussian) density functions reasonably well [148].

While orthogonal transformations between different domains are invertible in $\mathbb{R}^n$, the respective inverse transformation recovers the original values only approximately if the intermediate coefficients are rounded to fixed precision.[21] Embedding in the transformed domain, after possible rounding, is beneficial if this domain is also used on the channel, because subtle embedding changes are not at risk of being altered by later rounding in a different domain. Nevertheless, some stego systems intentionally choose a different embedding domain, and ensure robustness to later rounding errors with appropriate channel coding (e.g., embedding function YASS [218]).

In many lossy compression algorithms, different subbands are rescaled before rounding to reflect differences in perceptual sensitivity. Such scaling and subsequent rounding is called *quantisation*, and the scaling factors are referred to as *quantisation factors*. To ensure that embedding changes are not corrupted during quantisation, the embedding function is best applied on already quantised coefficients.

## 2.6.4 Selected Cover Formats: JPEG and MP3

In this section we review two specific cover formats, JPEG still images and MP3 audio, which are important for the specific results in Part II. Both formats are very popular (this is why they are suitable for steganography) and employ lossy compression to minimise file sizes while preserving good perceptual quality.

### 2.6.4.1 Essentials of JPEG Still Image Compression

The Joint Photographic Expert Group (JPEG) was established in 1986 with the objective to develop digital compression standards for continuous-tone still images, which resulted in ISO Standard 10918-1 [112, 183].

---

[21] This does not apply to the class of invertible integer approximations to popular transformations, such as (approximate) integer DCT and integer DWT; see, for example, [196].

Standard JPEG compression cuts a greyscale image into blocks of $8 \times 8$ pixels, which are separately transformed into the frequency domain by a 2D DCT. The resulting 64 DCT coefficients are divided by subband-specific quantisation factors, calculated from a JPEG quality parameter $q$, and then rounded to the closest integer. In the notation of Sect. 2.6.3, the quantised DCT coefficients $\boldsymbol{y}_{\boxplus}^{*}$ can be obtained as follows:

$$\boldsymbol{y}_{\boxplus}^{*} = \lfloor \bar{\boldsymbol{q}} \, \boldsymbol{y}_{\boxplus} + 1/2 \rfloor \quad \text{with} \quad \bar{q}_{i,j} = \begin{cases} (\mathsf{Quant}(q,i))^{-1} & \text{for } i = j \\ 0 & \text{otherwise.} \end{cases} \tag{2.7}$$

Function $\mathsf{Quant} : \mathbb{Z}^{+} \times \{1, \ldots, 64\} \rightarrow \mathbb{Z}^{+}$ is publicly known and calculates subband-specific quantisation factors for a given JPEG compression quality $q$. The collection of 64 quantisation factors on the diagonal of $\boldsymbol{q}$ is often referred to as *quantisation matrix* (then aligned to dimensions $8 \times 8$). In general, higher frequency subbands are quantised with larger factors. Then, the already quantised coefficients are reordered in a zigzag manner (to cluster 0s in the high-frequency subbands) and further compressed by a lossless run-length and Huffman entropy [107] encoder. A block diagram of the JPEG compression process is depicted in Fig. 2.9.

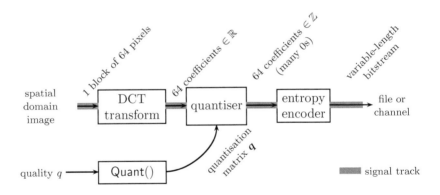

Fig. 2.9: Signal flow of JPEG compression (for a single colour component)

Colour images are first decomposed into a luminance component $\boldsymbol{y}$ (which is treated as a greyscale image) and two chrominance components $\boldsymbol{c}_R$ and $\boldsymbol{c}_B$ in the YCrCb colour model. The resolution of the chrominance components is usually reduced by factor 2 (owing to the reduced perceptibility of small colour differences of the human visual system) and then compressed separately in the same way as the luminance component. In general, the chrominance components are quantised with larger factors than the luminance component.

All JPEG operations in Part II were conducted with `libjpeg`, the Independent JPEG Group's reference implementation [111], using default settings for the DCT method unless otherwise stated.


### 2.6.4.2 Essentials of MP3 Audio Compression

The Moving Picture Expert Group (MPEG) was formed in 1988 to produce standards for coded representations of digital audio and video. The popular MP3 file format for lossy compressed audio signals is specified in the ISO/MPEG 1 Audio Layer-3 standard [113]. A more scientific reference is the article by Brandenburg and Stoll [30].

The MP3 standard combines several techniques to maximise the trade-off between perceived audio quality and storage volume. Its main difference from many earlier and less efficient compression methods is its design as a two-track approach. The *first track* conveys the audio information, which is first passed to a filter bank and decomposed into 32 equally spaced frequency subbands. These components are separately transformed to the frequency domain with a *modulated discrete cosine transformation* (MDCT).[22] A subsequent quantisation operation reduces the precision of the MDCT coefficients. Note that the quantisation factors are called 'scale factors' in MP3 terminology. Unlike for JPEG compression, these factors are not constant over the entire stream. Finally, lossless entropy encoding of the quantised coefficients ensures a compact representation of MP3 audio data. The *second track* is a control track. Also, starting again from the *pulse code modulation* (PCM) input signal, a 1024-point FFT is used to feed the frequency spectrum of a short window in time as input to a psycho-acoustic model. This model emulates the particularities of human auditory perception, measures and values distortion, and derives masking functions for the input signal to cancel inaudible frequencies. The model controls the choice of block types and frequency band-specific scale factors in the first track. All in all, the two-track approach adaptively finds an optimal trade-off between data reduction and audible degradation for a given input signal. Figure 2.10 visualises the signal flow during MP3 compression.

Regarding the underlying data format, an MP3 stream consists of a series of *frames*. Synchronisation tags separate MP3 audio frames from other information sharing the same transmission or storage stream (e.g., video frames). For a given bit rate, all MP3 frames have a fixed compressed size and represent a fixed amount of 1,152 PCM samples. Usually, an MP3 frame contains 32 bits of header information, an optional 16 bit *cyclic redundancy check*

---

[22] The MDCT corresponds to the *modulated lapped transformation* (MLT), which transforms overlapping blocks to the frequency domain [165]. This reduces the formation of audible artefacts at block borders. The inverse transformation is accomplished in an overlap-add process.

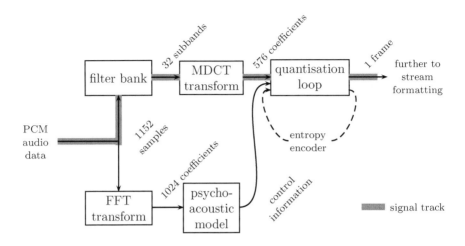

Fig. 2.10: Signal and control flow of MP3 compression (simplified)

(CRC) checksum, and two so-called *granules* of compressed audio data. Each granule contains one or two *blocks*, for mono and stereo signals, respectively. Both granules in a frame may share (part of) the scale factor information to economise on storage space. Since the actual block size depends on the amount of information that is required to describe the input signal, block and granule sizes may vary between frames. To balance the floating granule sizes across frames of fixed sizes efficiently, the MP3 standard introduces a so-called *reservoir* mechanism. Frames that do not use their full capacity are filled up (partly) with block data of subsequent frames. This method ensures that local highly dynamic sections in the input stream can be stored with over-average precision, while less demanding sections allocate under-average space. However, the extent of reservoir usage is limited in order to decrease the interdependencies between more distant frames and to facilitate resynchro-nisation at arbitrary positions in a stream. A schema of the granule-to-frame allocation in MP3 streams is depicted in Fig. 2.11.

### 2.6.5 Exotic Covers

Although the large majority of publications on steganography and ste-ganalysis deal with digital representations of continuous signals as covers,

variable-length granules

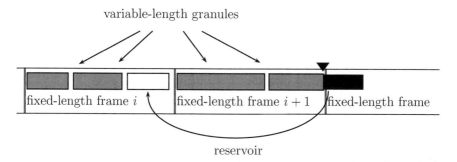

reservoir

Fig. 2.11: MP3 stream format and reservoir mechanism

alternatives have been explored as well. We mention the most important ones only briefly.

*Linguistic* or *natural language* steganography hides secret messages in text corpuses. A recent literature survey [13] concludes that this branch of research is still in its infancy. This is somewhat surprising as text covers have been studied in the very early publications on mimic functions by Wayner [232], and various approaches (e.g., lexical, syntactic, ontologic or statistical methods) of automatic text processing are well researched in computer linguistics and machine translation [93].

Vector objects, meshes and general graph-structured data constitute another class of potential covers. Although we are not aware of specific proposals for steganographic applications, it is well conceivable to adapt principles from watermarking algorithms and increase (steganographic) security at the cost of reduced robustness for steganographic applications. Watermarking algorithms have been proposed for a large variety of host data, such as 2D vector data in digital maps [136], 3D meshes [11], CAD data [205], and even for very general data structures, such as XML documents and relational databases [92]. (We cite early references of each branch, not the latest refinements.)

## 2.7 Embedding Operations

In an attempt to give a modular presentation of design options for steganographic systems, we distinguish the high-level embedding function from low-level *embedding operations*.

Although in principle Embed may be an arbitrary function, in steganography it is almost universal practice to decompose the cover into *samples* and the secret message into bits (or $q$-ary symbols), and embed bits (or symbols) into samples independently. There are various reasons for this being so popular: ease of embedding and extracting, ability to use coding methods, and ease of spreading the secret message over the cover. In the general set-ting, the assignment of message bits $m_j \in \{0,1\}$ to cover samples $x_i^{(0)}$ can be interleaved [43, 167]. Unless otherwise stated, we assume a pseudorandom permutation of samples using key $\boldsymbol{k}$ for secret-key steganography, although we abstract from this detail in our notation to improve readability. For em-bedding rates $p < 1$, random interleaving adds extra security by distributing the embedding positions over the entire cover, thus balancing embedding density and leaving the steganalyst uninformed about which samples have been changed for embedding (in a probabilistic sense). Below, in Sect. 2.8.2, we discuss alternative generalised interleaving methods that employ channel coding. These techniques allow us to minimise the number of changes, or to direct changes to specific parts of $\boldsymbol{x}^{(0)}$, the location of which remains a secret of the sender.

### 2.7.1 LSB Replacement

*Least significant bit* (LSB) replacement is probably the oldest embedding operation in digital steganography. It is based on the rationale that the right-most (i.e., least significant) bit in digitised signals is so noisy that its bitplane can be replaced by a secret message imperceptibly:

$$x_i^{(1)} \leftarrow 2 \cdot \lfloor x_i^{(0)}/2 \rfloor + m_j. \tag{2.8}$$

For instance, Fig. 2.12 shows an example greyscale image and its (ampli-fied) signal of the spatial domain LSB plane. The LSB plane looks purely random and is thus indistinguishable from the LSB plane of a stegotext with 12.5% secret message content. However, this impression is mislead-ing as LSBs, despite being superficially noisy, are generally not indepen-dent of higher bitplanes. This empirical fact has led to a string of powerful detectors for LSB replacement in the spatial domain [46, 48, 50, 73, 74, 82, 118, 122, 126, 133, 151, 160, 238, 252, 257] and in the DCT domain [152, 153, 238, 243, 244, 248, 251]. Note that some implementations of LSB replacement in the transformed domain skip coefficients with values $x^{(0)} \in \{0, +1\}$ to prevent perceptible artefacts from altering many 0s to val-ues +1 (0s occur most frequently due to the unimodal distribution with 0 mode). For the same reason, other implementations exclude $x^{(0)} = 0$ and modify the embedding function to

$$x_i^{(1)} \leftarrow 2 \cdot \left\lfloor (x_i^{(0)} - k)/2 \right\rfloor + k + m_j \quad \text{with} \quad k = \begin{cases} 0 \text{ for } x_i^{(0)} < 0 \\ 1 \text{ for } x_i^{(0)} > 0. \end{cases} \tag{2.9}$$

Probably the shortest implementation of spatial domain LSB replacement steganography is a single line of PERL proposed by Ker [118, p. 99]:

Fig. 2.12: Example eight-bit greyscale image taken from a digital camera and downsampled with nearest neighbour interpolation (left) and its least significant bitplane (right)

```
perl -n0777e '$_=unpack"b*",$_;split/(\s+)/,<STDIN>,5;
              @_[8]=~s{.}{$&&v254|chop()&v1}ge;
              print@_' <input.pgm >output.pgm secrettextfile
```

The simplicity of the embedding operation is often named as a reason for its practical relevance despite its comparative insecurity. Miscreants, such as corporate insiders, terrorists or criminals, may resort to manually typed LSB replacement because they must fear that their computers are monitored so that programs for more elaborate and secure embedding techniques are suspicious or risk detection as malware by intrusion detection systems (IDSs) [118].

### 2.7.2 LSB Matching (±1)

LSB matching, first proposed by Sharp [214], is almost as simple to implement as LSB replacement, but much more difficult to detect in spatial domain images [121]. In contrast to LSB replacement, in which even values are never decremented and odd values never incremented,[23] LSB matching chooses the change for each sample $x_i$ independently of its parity (and sign), for example, by randomising the sign of the change,

$$x_i^{(1)} \leftarrow x_i^{(0)} + \mathsf{LSB}(x_i^{(0)} - m_j) \cdot R_i \quad \text{with} \quad \frac{R_i + 1}{2} \sim \ddot{\mathcal{U}}_0^1. \tag{2.10}$$

Function $\mathsf{LSB} : \mathcal{X} \to \{0, 1\}$ returns the least significant bit of its argument,

---

[23] This statement ignores other conditions, such as in Eq. (2.9), which complicate the rule but do not solve the problem of LSB replacement that the steganalyst can infer the sign of potential embedding changes.

$$\mathsf{LSB}(x) = x - 2 \cdot \lfloor x/2 \rfloor = \mathsf{Mod}(x, 2). \tag{2.11}$$

$R_i$ is a discrete random variable with two possible realisations $\{-1, +1\}$ that each occur with 50% probability. This is why LSB matching is also known as $\pm 1$ embedding ('plus-minus-one', also abbreviated PM1). The random signs of the embedding changes avoid structural dependencies between the direction of change and the parity of the sample, which defeats those detection strategies that made LSB replacement very vulnerable. Nevertheless, LSB matching preserves all other desirable properties of LSB replacement. Message extraction, for example, works exactly in the same way as before: the recipient just interprets $\mathsf{LSB}(x_i^{(1)})$ as message bits.

If Eq. (2.10) is applied strictly, then elements $x_i^{(1)}$ may exceed the domain of $\mathcal{X}$ if $x_i^{(0)}$ is saturated.[24] To correct for this, $\boldsymbol{R}$ is adjusted as follows: $R_i = +1$ for $x_i^{(0)} = \inf \mathcal{X}$, and $R_i = -1$ for $x_i^{(0)} = \sup \mathcal{X}$. This does not affect the steganographic semantic for the recipient, but LSB matching reduces to LSB replacement for saturated pixels. This is why LSB matching is not as secure in covers with large areas of saturation. A very short PERL implementation for random LSB matching is given in [121].

Several variants of embedding functions based on LSB matching have been proposed in the literature and shall be recalled briefly:

- **Embedding changes with moderated sign** If reasonably good distribution models are known for cover signals, then the sign of $R_i$ can be chosen based on these models to avoid atypical deformation of the histogram. In particular, $R_i$ should take value $+1$ with higher probability in regions where the density function has a positive first derivative, whereas $R_i = -1$ is preferable if the first derivative of the density function is negative. For example, the F5 algorithm [233] defines fixed signs of $R_i$ depending on which side of the theoretical (0 mean) distribution of quantised JPEG AC coefficients a realisation $x_i^{(0)}$ is located. Hence, it embeds bits into coefficients by never increasing their absolute value.[25] Possible ambiguities in the steganographic semantic for the recipient can be dealt with by re-embedding (which gives rise to the 'shrinkage' phenomenon: for instance, algorithm F5 changes 50% of $x_i^{(0)} \in \{-1, +1\}$ without embedding a message bit [233]), or preferably by suitable encoding to avoid such cases preemptively (cf. Sect. 2.8.2 below).

---

[24] Saturation means that the original signal went beyond the bounds of $\mathcal{X}$. The resulting samples are set to extreme values $\inf \mathcal{X}$ or $\sup \mathcal{X}$.

[25] Interestingly, while this embedding operation creates a bias towards 0 and thus changes the shape of the histogram, Fridrich and Kodowsky [86] have proven that this operation introduces the least overall embedding distortion if the unquantised coefficients are unknown (i.e., if the cover is already JPEG-compressed). This finding also highlights that small distortion and histogram preservation are competing objectives, which cannot be optimised at the same time.

- **Determining the sign of $R_i$ from side information** Side informa-
  tion is additional information about the cover $\boldsymbol{x}^{(0)}$ available *exclusively*
  to the sender, whereas moderated sign embedding uses global rules or
  information shared with the communication partners. In this sense, side
  information gives the sender an advantage which can be exploited in the
  embedding function to improve undetectability. It is typically available
  when Embed goes along with information loss, for example, through scale
  reduction, bit-depth conversions [91], or JPEG (double-)compression [83]
  (cf. Fig. 2.4 in Sect. 2.4.2, where the lossy operation is explicit in function
  Process). In all these cases, $\boldsymbol{x}^{(0)}$ is available at high (real) precision and
  later rounded to lower (integer) precision. If $R_i$ is set to the opposite sign
  of the rounding error, a technique known as *perturbed quantisation* (PQ),
  then the total distortion of rounding and embedding decreases relative
  to the independent case, because embedding changes always offset a fraction of the rounding error (otherwise, the square errors of both distortions
  are additive, a corollary of the theorem on sums of independent random
  variables). Less distortion is believed to result in less detectable stego objects, though this assumption is hard to prove in general, and pathologic
  counterexamples are easy to find.
- **Ternary symbols: determining the sign of $R_i$ from the secret message** The direction of the change can also be used to convey additional
  information if samples of $\boldsymbol{x}^{(1)}$ are interpreted as ternary symbols (i.e., as
  representatives of $\mathbb{Z}_3$) [169]. In a fully ternary framework, a net capacity
  of $\log_2 3 \approx 1.585$ bits per cover symbol is achievable, though it comes at
  a cost of potentially higher detectabiliy because now $2/3$ of the symbols
  have to be changed on average, instead of $1/2$ in the binary case (always assuming maximum embedding rates) [91]. A compromise that uses ternary
  symbols to embed one extra bit per block—the operation is combined with
  block codes—while maintaining the average fraction of changed symbols at
  $1/2$ has been proposed by Zhang et al. [254]. Ternary symbols also require
  some extra effort to deal with $x_i^{(0)}$ at the margins of domain $\mathcal{X}$.

All embedding operations discussed so far have in common the property
that the maximal absolute difference between individual cover symbols $x_i^{(0)}$
and their respective stego symbols $x_i^{(1)}$ is $1 \geq |x_i^{(0)} - x_i^{(1)}|$. In other words,
the maximal absolute difference is minimal. A visual comparison of the similarities and differences of the mapping between cover and stego samples is
provided in Fig. 2.13 (p. 44).

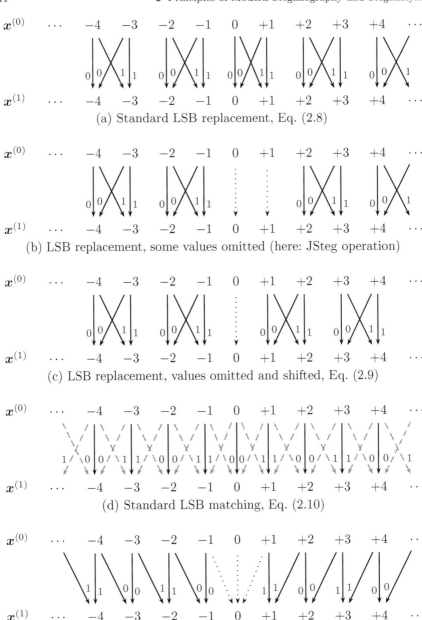

Fig. 2.13: Options for embedding operations with minimal maximum absolute embedding distortion per sample: $\max |x_i^{(0)} - x_i^{(1)}| = 1$; dotted arrows represent omitted samples, dashed arrows are options taken with conditional probability below 1 (condition on the message bit); arrow labels indicate steganographic semantic after embedding

## 2.7.3 Mod-k Replacement, Mod-k Matching, and Generalisations

If stronger embedding distortions $|x_i^{(0)} - x_i^{(1)}|$ than 1 are acceptable, then embedding operations based on both replacement and matching can be generalised to larger alphabets by dividing domain $\mathcal{X}$ into $N$ disjoint sets of subsequent values $\{\mathcal{X}_i \mid \mathcal{X}_i \subset \mathcal{X} \wedge |\mathcal{X}_i| \geq k, 1 \leq i \leq N\}$. The steganographic semantic of each of the $k$ symbols in the (appropriately chosen) message alphabet can be assigned to exactly one element of each subset $\mathcal{X}_i$. Such subsets are also referred to as *low-precision bins* [206].

For $\mathbb{Z}_{Nk} \subset \mathcal{X}$, a suitable breakdown is $\mathcal{X}_i = \{x \mid \lfloor x/k \rfloor = i - 1\}$ so that each $\mathcal{X}_i$ contains distinct representatives of $\mathbb{Z}_k$. The $k$ symbols of the message alphabet are assigned to values of $x^{(1)}$ so that $\mathsf{Mod}(x^{(1)}, k) = m$. Mod-$k$ replacement maintains the low-precision bin after embedding (hence $x^{(0)}, x^{(1)} \in \mathcal{X}_i$) and sets

$$x_i^{(1)} \leftarrow k \cdot \lfloor x_i^{(0)}/k \rfloor + m_j. \tag{2.12}$$

For $k = 2^z$ with $z$ integer, mod-$k$ replacements corresponds to LSB replacement in the $z$ least significant bitplanes.

Mod-$k$ matching picks representatives of $m_j \equiv x_i^{(1)} (\bmod\ k)$ so that the embedding distortion $|x^{(0)} - x^{(1)}|$ is minimal (random assignment can be used if two suitable representatives are equally distant from the cover symbol $x^{(0)}$).

Further generalisations are possible if the low-precision bins have different cardinalities, for example, reflecting different tolerable embedding distortions in different regions of $\mathcal{X}$. Then, the message has to be encoded to a mixed alphabet. Another option is the adjustment of marginal symbol probabilities using *mimic functions*, a concept introduced by Wayner [232]. Sallee [206] proposed arithmetic decoders [240] as tools to build mimic functions that allow the adjustment of symbol probabilities in mod-$k$ replacement conditionally on the low-precision bin of $x^{(0)}$.

Figure 2.14 illustrates the analogy between source coding techniques and mimic functions: in traditional source coding, function Encode compresses a nonuniformly distributed sequence of source symbols into a, on average, shorter sequence of uniform symbol distribution. The original sequence can be recovered by Decode with side information about the source distribution. Mimic functions useful in steganography can be created by swapping the order of calls to Encode and Decode: a uniform message sequence can be transcoded by Decode to an exogenous target distribution (most likely to match or 'mimic' some statistical property of the cover), whereas Encode is called at the recipient's side to obtain the (uniform, encrypted) secret message sequence.

*Stochastic modulation embedding* [72] is yet another generalisation of mod-$k$ matching which allows (almost) arbitrary distribution functions for the

**Source coding**

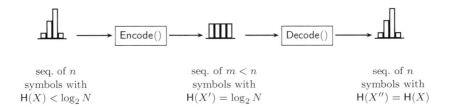

seq. of $n$                    seq. of $m < n$                    seq. of $n$
symbols with               symbols with                    symbols with
$H(X) < \log_2 N$          $H(X') = \log_2 N$              $H(X'') = H(X)$

**Mimic function**

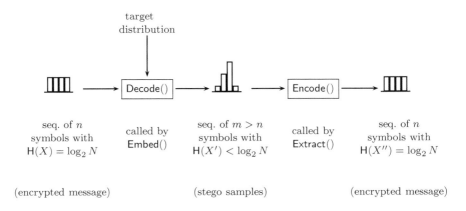

seq. of $n$          called by          seq. of $m > n$          called by          seq. of $n$
symbols with       Embed()           symbols with            Extract()         symbols with
$H(X) = \log_2 N$                     $H(X') < \log_2 N$                          $H(X'') = \log_2 N$

(encrypted message)                   (stego samples)                   (encrypted message)

Fig. 2.14: Application of source coding techniques for entropy encoding (top) and as mimic function for embedding (bottom). The alphabet size is $N$ and input sequences are identical to output sequences in both cases

random variable $R$ in Eq. (2.10). The sender uses a pseudorandom number generator (PRNG) with a seed derived from the secret key to draw realisations from $R_i$. This ensures that the recipient can reproduce the actual sequence of $r_i$ and determine the positions of samples where $|r_i|$ is large enough so that both steganographic message bits could be embedded by either adding or subtracting $r_i$ from $x_i^{(0)}$ to obtain $x_i^{(1)}$. Extract evaluates only these 'usable' positions while skipping all others.

Finally, *spread spectrum image steganography* (SSIS) [167] can be seen as an approximate version of stochastic modulation (though invented before) which does not preemptively skip unusable realisations of $R_i$. To achieve comparable embedding capacities, on average higher embedding distortions have to be accepted, which require extra redundancy through error correction codes and signal restoration techniques on the recipient's side. However, this extra effort lends SSIS a slight advantage over pure stochastic modulation in terms of robustness. SSIS, despite its name, is not limited to images as cover.

### 2.7.4 Multi-Sample Rules

As it is difficult to ensure that samples can be modified independently without leaving detectable traces, multi-sample rules have been proposed to change samples $x_i^{(0)}$ conditional on the realisations of other samples $x_j^{(0)}, j \neq i$, or even jointly. We distinguish broadly between two kinds of reference samples:

- Reference samples $x_j^{(0)}$ can be located in either spatial or temporal proximity, where the dependencies are assumed to be stronger than between more distant samples.
- Aggregate information of all samples in a cover object can serve as reference information. The idea here is to preserve macroscopic statistics of the cover.

One example for the first kind is the embedding operation of the CAS scheme by Lou and Sung [159], which evaluates the average intensity of the top-left adjacent pixels as well as the bottom-right adjacent pixels to calculate the intensity of the centre pixel conditional on the (encrypted) message bit (we omit the details for brevity). However, the CAS scheme shares a problem of multi-sample rules which, if not carefully designed, often ignore the possibility that a steganalyst who knows the embedding relations between samples can count the number of occurrences in which these relation hold exactly. This information, possibly combined with an analysis of the distribution of the exact matches, is enough to successfully detect the existence of hidden messages [21]. Another caveat of this kind of multi-sample rule is the need to ensure that subsequent embedding changes to the reference samples do not wreck the recipient's ability to identify the embedding positions (i.e., the criterion should be invariant to embedding operations on the reference samples).

Pixel-value differencing (PVD) in spatial domain images is another example of the first kind. Here, mod-$k$ replacement is applied to intensity differences between pairs [241] or tuples [39] of neighbouring samples, possibly combined with other embedding operations on intensity levels or compensation rules to avoid unacceptable visible distortion [242]. Zhang and Wang [256] have proposed a targeted detector for PVD.

Examples for the second kind of multi-sample rules are OutGuess by Provos [198] and StegHide by Hetzl and Mutzel [102]. OutGuess employs LSB replacement in JPEG DCT coefficients, but flips additional correction LSBs to preserve the marginal histogram distributions. This increases the average distortion per message bit and makes the stego system less secure against all kinds of detectors which do not only rely on marginal distributions. For instance, the detector by Fridrich et al. [76], calculates blockiness measures in the spatial domain. StegHide [102] preserves marginal distributions of arbitrary covers by exchanging positions of elements in $\boldsymbol{x}^{(0)}$ rather than altering values independently. A combinatorial solution is found by expressing the relations for possible exchanges as edges of a (possibly weighted) graph, which is solved by maximum cardinality matching. Successful steganalysis of StegHide has been reported for audio [204] and JPEG [157] covers. Both detectors evaluate statistics beyond the preserved marginal distributions.

## 2.7.5 Adaptive Embedding

Adaptive embedding can be seen as a special case of multi-sample rules; however, information from reference samples is not primarily used to apply consistent changes, but rather to identify locations where the distortion of single-sample embedding operations is least detectable. The aim is to concentrate the bulk of necessary changes there. Adaptive embedding can be combined with most of the above-discussed embedding operations. Ideally, the probability that the embedding operation does not modify a particular sample value should be proportional to the information advantage of the steganalyst from observing this particular sample in a modified realisation[26]:

$$\mathsf{Prob}(x_i^{(1)} = x_i^{(0)}) \propto \mathsf{Prob}(j \neq 0 | \boldsymbol{x}^{(j)} \wedge x_i^{(j)} \neq x_i^{(0)}) - \mathsf{Prob}(j \neq 0 | \boldsymbol{x}^{(j)}). \quad (2.13)$$

Unfortunately, the probabilities on the right-hand side of this relation are unknown in general (unless specific and unrealistic assumptions for the cover are made). Nevertheless, heuristic proposals for adaptive embedding rules are abundant for image steganography.[27] Lie and Chang [154] employ a model of the human visual system to control the number $k$ of LSB planes used for mod-$2^k$ replacement. Franz, Jerichow, Möller, Pfitzmann, and Stierand [63] exclude values close to saturation and close to the zero crossing of PCM digitised speech signals. Franz [62] excludes entire histogram bins from embedding based on the joint distribution with adjacent bins in a co-occurrence matrix built from spatial relations between pixels. Fridrich and Goljan [72]

---

[26] Note that this formulation states adaptive steganography as a local problem. Even if it could be solved for each sample individually, the solution would not necessarily be optimal on a global (i.e., cover-wide) scope. This is so because the individual information advantage may depend on other samples' realisations. In this sense, Eq. (2.13) is slightly imprecise.

[27] Despite the topical title 'Adaptive Steganography' and some (in our opinion) improper citations in the context of adaptive embedding operations, reference [37] does not deal with adaptive steganography according to this terminology. The paper uses adaptive in the sense of anticipating the steganalyst's exact detection method, which we deem rather unrealistic for security considerations.

discuss a content-dependent variant of their stochastic modulation operation, in which the standard deviation of the random variable $R$ is modulated by an energy measure in the spatial neighbourhood. Similarly, adaptive ternary LSB matching is benchmarked against various other embedding operations in [91]. Aside from energy measures, typical image processing operators were suggested for adaptive steganography, such as dithering [66], texture [101] and edge detectors [180, 241, 242, 245].[28] Probably the simplest approach to adaptive steganography is due to Arjun et al. [6], who use the assumed perceptibility of intensity difference depending on the magnitude of $x_i^{(0)}$ as criterion, complemented by an exclusion of pixels with a constant intensity neighbourhood.

At first sight, adaptive embedding appears beneficial for the security of a stego system independent of the cover representation or embedding function [226] (at least if the underlying embedding operation is not insecure per se; so avoid LSB replacement). However, this only helps against myopic adversaries: one has to bear in mind that many of the adaptivity criteria are (approximately) invariant to embedding. In some embedding functions this is even a requirement to ensure correct extraction.[29] Adhering to Kerckhoffs' principle [135], this means that the steganalyst can re-recognise those regions where embedding changes are more concentrated. And in the worst case, the steganalyst could even compare statistics between the subset of samples which might have been affected from embedding and others that are most likely in their original state. Such kinds of detectors have been demonstrated against specific stego systems, for example, in [24]. More general implications of the game between steganographers and steganalysts on where to hide (and where to search, respectively) are largely unexplored. One reason for this gap might be the difficulty of quantifying the *detectability profile* [69] as a function of general cover properties. In Chapter 5 we present a method which is generally suitable to estimate cost functions for covers (and individual pixels, though not part of this book) empirically.

## 2.8 Protocols and Message Coding

This section deals with the architecture of stego systems on a more abstract level than the actual embedding operation on the signal processing layer. Topics of interest include the protocol layer, in particular assumptions on key distribution (Sect. 2.8.1), and options for coding the secret message to

---

[28] All these references evaluate the difference between neighbouring pixels to adjust $k$ in mod-$k$ replacement of the sample value or pairwise sample differences (i.e., PVDs). They differ in the exact calculation and correction rules to ensure that Extract works.

[29] Wet paper codes (cf. 2.8.2.2) have proved a recipe for correct extraction despite keeping the exact embedding positions a secret.

minimise the (detectability-weighted) distortion or leverage information advantages of the sender over the steganalyst (coding layer, Sect. 2.8.2).


## 2.8.1 Public-Key Steganography

In the context of steganography, the role of cryptography and of cryptographic keys in particular is to distinguish the communication partners from the rest of the world. Authorised recipients are allowed to recognise steganographic content and even extract it correctly, whereas third parties must not be able to tell stego objects apart from other communications. The common assumption in Simmons' initial formulation of the prisoners' problem [217] is that both communication partners share a common secret. This implies that both must have had the opportunity to communicate securely in the past to agree on a symmetric steganographic key. Moreover, they must have anticipated a situation in which steganographic communication is needed.[30]

Cryptography offers ways to circumvent this key distribution problem by using *asymmetric* cryptographic functions that operate with pairs of public and private keys. There exist no proposals like 'asymmetric steganography' for a direct analogy in steganography. Such a construction would require a trapdoor embedding function that is not invertible without the knowledge of a secret (or vast computational resources). However, by combining asymmetric cryptography with symmetric embedding functions, it is possible to construct so-called *public-key steganographic systems* (acronym PKS, as opposed to SKS for *secret-key steganography*).

The first proposal of steganography with public keys goes back to Anderson's talk on the first Information Hiding Workshop in 1996 [4]. Since, his work has been extended by more detailed considerations of active warden models [5]. The construction principles are visualised as a block diagram in Fig. 2.15, where we assume a passive warden adversary model. The secret message is encrypted with the public key of the recipient using an asymmetric cryptographic function, then (optionally) encoded so that encrypted message bits can be adapted to marginal distributions of the cover (mimic function) or placed in the least conspicuous positions in the cover. A keyless embedding function finally performs the actual embedding.[31] The recipient extracts a bitstream from each received object, feeds it to the decoder and subsequently tries to decrypt it with his or her private key. If the decryption

---

[30] It is obvious that allowing secret key exchanges in general when already 'locked in Simmons' prison' would weaken the assumptions on the communication restrictions: communication partners who are allowed to exchange keys (basically random numbers) can communicate *anything* through this channel.

[31] For example, a symmetric embedding function suitable for SKS with globally fixed key $k = \text{const}$.

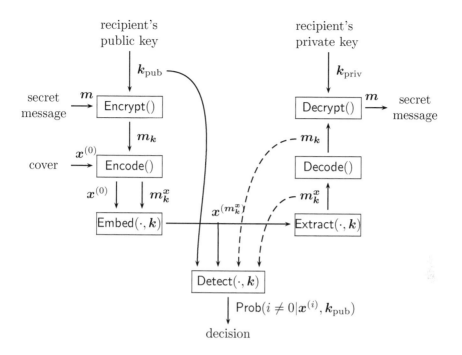

Fig. 2.15: Block diagram of public-key stego system with passive warden. Dashed lines denote that the information can be derived from $x^{(m_k^x)}$ with public knowledge. The global 'key' $k$ is optional and can be part of Embed, Extract and Detect (Kerckhoffs' principle)

succeeds, the recipient recognises that the received object was actually a stego object and retrieves the secret message.[32]

It is obvious that such a PKS system can never be more secure than the underlying SKS stego system consisting of Embed and Extract for random messages of length $|m|$. In addition, as can be seen from the high number of arrows pointing to the steganalyst's function Detect, it is important for the security of the construction that none of

- the stego object $x^{(m_k^x)}$,
- the bitstream generated by the message encoder $m_k^x$, and
- the encrypted message $m_k$

be statistically distinguishable between clean covers and stego objects, even with knowledge of the recipient's public key $k_{\mathrm{pub}}$ (and, if it exists, knowledge

---

[32] Note that the message coding is implicit as part of Embed in the original publication. The distinction is made in the figure to emphasise which components of the system must output information indistinguishable between clean covers and stego objects [16].

of the global 'key' $k$ used in the symmetric stego system Embed and Extract). In other words, Extract applied to arbitrary covers must always return a random sequence (possibly correlated to $x$, but never revealing that information about $x^{(0)}$ if $x^{(p)}$ with $p > 0$ has been transmitted). Moreover, Decode applied to any possible output of Extract should be indistinguishable from ciphertexts created with Encrypt and the recipient's public key $k_{pub}$. Only few asymmetric encryption schemes produce pseudorandom ciphertexts (e.g., [171] for a scheme based on elliptic curves, which has the nice property that it produces shorter ciphertexts than RSA or Diffie–Hellman-based alternatives), and well-known number-theoretic schemes in $\mathbb{Z}_p$ or $\mathbb{Z}_n$, with $p$ prime and $n$ semi-prime, can be used for PKS only in conjunction with a *probabilistic bias removal* (PBR) procedure [246].[33]

Initiating a steganographic communication relation with public keys requires a key exchange protocol, in which the recipient transmits his or her public key to the sender (and thus, at the same time, reveals his or her intention to communicate covertly). Assuming that sending keys openly is considered as suspicious, the public key itself has to be embedded as a secret message [44]. Again, one has to ensure that public keys are pseudorandom, which is not the case for the RSA-based key exchange proposed by Craver [44] (because random numbers tend to have small factors, but the semi-prime $n$ part of the RSA public key does not).[34] Therefore, a Diffie–Hellman integer encryption scheme (DHIES) [2] augmented by a PBR for the key exchange should be sufficiently secure in the passive warden model (NB, against polynomial bounded adversaries; if secure SKS exists; if the hash and MAC functions in the concrete DHIES implementation are secure).

Steganographic key exchanges are yet more difficult in the active warden adversary model. As discussed before in Sect. 2.5.2 (p. 29), we are not aware of a solution to the 'stego challenge' problem. A different approach to completely avoid the troublesome key exchanges in PKS is the (convenient) assumption that all communication partners have access to a digital signature system and can reuse its keys for steganography [144].

Orthogonal to extensions of Anderson's construction [4, 21, 44, 94, 144], there are publications on public-key steganography originating from the cryptology community. This literature focuses on public-key steganographic systems with provable security properties even in active warden models [3, 8, 104, 150]. However, the cost of this formal rigour is practical irrelevance, essentially due to two constraints, namely unrealistic assumptions,

---

[33] This is so because valid ciphertexts $s < n$, but $\lceil \log_2 n \rceil$ bits are needed to store $s$, so the distribution of 0s and 1s in the most significant bit(s) is not uniform.

[34] One can differentiate between whether it is sufficient that a notably high number of clean covers 'contain' a plausible public key, or whether finding a cover that does not 'contain' a message distinguishable from possible public keys should be difficult. While the former condition seems reasonable in practice, the latter is stronger and allows an admittedly unrealistic regime in which all complying communication partners who 'have nothing to hide' actively avoid sending covers with plausible public stego keys in order to signal their 'stegophobia', and thus potential steganographers are singled out.

most importantly that cover symbols can be sampled from an artificial channel with a known distribution, and inefficiency (such as capacities of one bit per cover object). The difference between these rather theoretical constructions of provable secure steganography and practical systems are not specific to PKS and explained further in Sect. 3.4.4.

## 2.8.2 Maximising Embedding Efficiency

Another string of research pioneered by Anderson [4] and, more specifically, Crandall [43] and Bierbrauer [14] includes channel coding techniques in the embedding function to optimise the choice of embedding positions for minimal detectability. As soon as the length of the secret message to be embedded $|m|$ is smaller than the number of symbols $n$ in $x^{(0)}$ (with binary steganographic semantic), the sender gains degrees of freedom on which symbols to change to embed $m$ in the least-detectable way, that is, with highest *embedding efficiency*. In general, embedding efficiency $\eta$ can be defined as the length of the secret message divided by a suitable distortion measure for the steganographic system and adversary model under consideration:

$$\eta = \frac{|m|}{\text{embedding distortion}}. \tag{2.14}$$

We distinguish between two important specific distortion measures, although other metrics and combinations are conceivable as well.

### 2.8.2.1 Embedding Efficiency with Respect to the Number of Changes

A simple measure of distortion is the number of changes to cover $x^{(0)}$ during embedding; hence, Eq. (2.14) can be written as

$$\eta_{\#} = \frac{|m|}{\mathsf{D_H}(x^{(0)}, x^{(m)})} \quad \text{with} \quad \mathsf{D_H}(x, y) = \sum_i (1 - \delta_{x_i y_i}). \tag{2.15}$$

Function $\mathsf{D_H} : \mathcal{X}^n \times \mathcal{X}^n \to \mathbb{Z}$ denotes the Hamming distance between two vectors of equal length. *Syndrome coding* is a technique borrowed from channel coding to improve $\eta_{\#}$ above a value of 2.[35] To cast our cover vectors (following optional key-dependent permutation) to the universe of block codes, we

---

[35] If bits in $m$ and the steganographic semantic of symbols in $x^{(0)}$ are independently distributed with maximum entropy, then on average one symbol has to be changed to embed two message bits (the steganographic semantic of cover symbols already matches the desired message bit with 50% chance).

interpret $\boldsymbol{x}^{(0)} = (x_1, \ldots, x_n) = \boldsymbol{x}_1^{(0)} || \boldsymbol{x}_2^{(0)} || \ldots || \boldsymbol{x}_{\lceil n/n_\square \rceil}^{(0)}$ as a concatenation of blocks of size $n_\square$ each. Let $\boldsymbol{d} \in \{0,1\}^*$ be an $l \times n_\square$ parity check matrix of a linear block code (with $\mathsf{rank}(\boldsymbol{d}) = l \leq n_\square$), and let $\boldsymbol{b}_j^{(0)} \in \{0,1\}^{n_\square}$ be the binary column vector of the steganographic semantic extracted from individual symbols of $\boldsymbol{x}_j^{(0)}$, the $j$th block of $\boldsymbol{x}^{(0)}$.

If the recipient, after extracting the steganographic semantic $\boldsymbol{b}_j^{(1)}$ from $\boldsymbol{x}_j^{(1)}$, always builds the matrix product

$$\boldsymbol{m}_j = \boldsymbol{d}\,\boldsymbol{b}_j^{(1)} \qquad (2.16)$$

to decode $l$ message bits $\boldsymbol{m}_j$, then the sender can rearrange Eq. (2.16) and search for the auxiliary vector $\boldsymbol{v}_j$ that solves Eq. (2.19) with minimal Hamming weight. Nonzero elements in $\boldsymbol{v}_j$ indicate $\mathsf{D_H}(\boldsymbol{v}, \boldsymbol{0})$ positions in $\boldsymbol{x}_j^{(0)}$ where the steganographic semantic has to be changed by applying the embedding operation,

$$\boldsymbol{v}_j = \boldsymbol{b}_j^{(1)} - \boldsymbol{b}_j^{(0)} \qquad (2.17)$$

$$\boldsymbol{d}\,\boldsymbol{v}_j = \boldsymbol{d}\,\boldsymbol{b}_j^{(1)} - \boldsymbol{d}\,\boldsymbol{b}_j^{(0)} \qquad (2.18)$$

$$\boldsymbol{d}\,\boldsymbol{v}_j = \boldsymbol{m}_j - \boldsymbol{d}\,\boldsymbol{b}_j^{(0)}. \qquad (2.19)$$

The syndrome $\boldsymbol{d}\,\boldsymbol{b}_j^{(0)}$ lends its name to the technique.

Early proposals [43] for the creation of $\boldsymbol{d}$ suggest binary Hamming and Golay codes, which are both good error-correcting codes and covering codes (the latter is important for embedding purposes). All codes of the Hamming family [96] are perfect codes and share a minimum distance 3 and a covering radius 1, which implies that the weight of $\boldsymbol{v}_j$ never exceeds 1. The only remaining perfect binary code is the binary Golay code, which has minimum distance 7 and covering radius 3 [14]. The advantage of Hamming codes is that the search for $\boldsymbol{v}_j$ is computationally easy—it follows immediately from the difference between syndrome $\boldsymbol{d}\,\boldsymbol{b}_j^{(0)}$ and message $\boldsymbol{m}_j$. This is why Hamming codes, renamed as 'matrix codes' in the steganography community, found their way into practical embedding functions quickly [233, for example]. More recently, covering properties of other structured error-correcting codes, such as BCH [173, 210, 211, 250], Reed–Solomon [61], or simplex (for $|\boldsymbol{m}|/n$ close to 1) [88], as well as (unstructured) random linear codes [85], have been studied.

A common problem of structured error-correcting codes beyond the limited set of perfect codes are their comparatively weak covering properties and the exponential complexity (in $n_\square$) of the search for $\boldsymbol{v}_j$ with minimum weight (also referred to as *coset leader* in the literature). This imposes an upper limit on possible block size $n_\square$ and keeps the attainable embedding efficiencies $\eta_\#$ in the low region of the theoretical bound [14]. Even so, heuristics have been proposed to trade off computational and memory complexity, to employ probabilistic processing, and to restrict the result set to approximate (local) solutions [71, 212]. More recent methods exploit structural properties of the code [250] or are based on *low-density generator matrix* (LDGM) codes. For the latter, approximate solutions can be found efficiently for very large $n_\square \approx n$ [71, 95]. LDGM solvers can handle weighted Hamming distances and seem to work with more general distortion measures (of which Sect. 2.8.2.2 is a special case).

Most coding techniques mentioned here are not limited to binary cases, and some generalisations to arbitrary finite fields exist (e.g., Bierbrauer [14] for the general theory, Willems and van Dijk [239] for ternary Hamming and Golay codes, Fridrich [69] for $q$-ary random codes on groups of binary samples, and Zhang et al. [255] for code concatenation of binary codes in 'layers').

### 2.8.2.2 Embedding Efficiency with Respect to the Severity of Changes

Consider a function that implements adaptive embedding (cf. Sect. 2.7.5), possibly taking into account additional side information,

$$\mathsf{Wet} : \mathcal{X}^n \times \{\mathbb{R}^n, \bot\} \to \{0, 1\}^n, \tag{2.20}$$

which assigns each sample in $\boldsymbol{x}^{(0)}$ to one of two classes based on the severity of a change with respect to perceptibility or detectability. Samples that are safe to be changed are called 'dry' (value 0) and those that should not be altered are called 'wet' (value 1). A useful metaphor is a piece of paper besprinkled in rain, so that ink lasts only on its dry parts. After a while, primarily 'wet' and 'dry' regions cannot be told apart anymore. This led to the term *wet paper codes* for embedding, introduced by Fridrich, Goljan, and Soukal [83].

Possible denominators of Eq. (2.14) can be arbitrary projections of the value of Wet to a scalar, such as the number of 'wet' samples changed; or, if the co-domain of Wet is continuous, a weighted sum. For the sake of simplicity, we restrict the presentation to this (degenerated, but fairly common) binary case:

$$\eta_\odot = \begin{cases} 1 \text{ for } & x_i^{(0)} = x_i^{(m)} \ \forall i \in \{i \mid \mathsf{Wet}(\boldsymbol{x}^{(0)}, \cdot) = 1\} \\ 0 \text{ otherwise.} \end{cases} \tag{2.21}$$

According to this definition, embedding is efficient if the message can be placed into the cover object without altering any 'wet' sample and the recipient is able to extract it correctly without knowing the value of Wet. A first proposal for this problem by Anderson [4] is known as *selection channel*: all elements of $\boldsymbol{x}^{(0)}$ are divided into $|\boldsymbol{m}| \ll n$ blocks $\boldsymbol{x}_1^{(0)} || \boldsymbol{x}_2^{(0)} || \dots || \boldsymbol{x}_{|\boldsymbol{m}|}^{(0)}$. Then, the parity of the steganographic semantics of all samples in one block is interpreted as a message bit. Only blocks for which the parity does not match the message bit, i.e., $m_i \neq \mathsf{Parity}(\boldsymbol{b}_i^{(0)})$, must be adjusted by selecting the least-detectable sample of $\boldsymbol{x}_i^{(0)}$ for the embedding operation. If $n/|\boldsymbol{m}|$ is sufficiently large and elements of $\boldsymbol{x}^{(0)}$ are assigned to blocks $\boldsymbol{x}_i^{(0)}$ randomly, then the probability that no 'wet' sample has to be changed is reasonably high.

The probability of successful embedding can be further improved by using *wet paper codes* (WPCs), a generalisation of the selection channel. As for the minimisation of the number of changes, block sizes $n_\square = |\boldsymbol{x}_i^{(0)}|$ are chosen larger (hundreds of samples) to accommodate $l$ message bits per block. For each block, an $l \times n_\square$ parity check matrix $\boldsymbol{d}_j$ is populated using a pseudorandom number generator seeded with key $\boldsymbol{k}$. As before, $\boldsymbol{b}_j^{(0)}$ is the steganographic semantic extracted from $\boldsymbol{x}_j^{(0)}$, and $\overline{\boldsymbol{b}}_j^{(0)}$ is a decimated vector excluding all bits that correspond to 'wet' samples. Analogously, the respective columns in $\boldsymbol{d}_j$ are removed in the reduced $l \times (n_\square - k_j)$ matrix $\overline{\boldsymbol{d}}_j$ ($k_j$ is the number of 'wet' samples in the $j$th block, and $n_\square - k_j \gtrsim l$). Vector $\overline{\boldsymbol{v}}_j$ indicates the embedding positions after inserting 0s for the omitted 'wet' samples and can be obtained by solving this equation with the Gaussian elimination method over the finite field $\mathbb{Z}_2$:[36]

$$\overline{\boldsymbol{d}}_j \, \overline{\boldsymbol{v}}_j = \boldsymbol{m}_j - \boldsymbol{d}_j \, \boldsymbol{b}_j^{(0)}. \tag{2.22}$$

As shown in [31] (cited from [83]), solutions for this system exist with high probability if $\boldsymbol{d}_j$ is sparsely populated. Unlike in the case of minimal changes, *any* solution is sufficient and there are no constraints with regard to the Hamming weight of $\overline{\boldsymbol{v}}_j$. The decoding operation is similar to Eq. (2.16) and uses the unreduced random matrix $\boldsymbol{d}_j$, since the recipient by definition does not know which columns were dropped due to 'wet' samples:

$$\boldsymbol{m}_j = \boldsymbol{d}_j \, \boldsymbol{b}_j^{(1)}. \tag{2.23}$$

Detailed strategies to embed the dimension of $\boldsymbol{d}$ (needed by the recipient) as metadata (obviously not using WPC) as well as a computationally less complex substitute for the Gaussian elimination, which exploits a specific stochastic structure of row and column weights in $\boldsymbol{d}_j$ and $\overline{\boldsymbol{d}}_j$, can be found in [80] and [81].

---

[36] Wet paper codes can be generalised to finite fields $\mathbb{Z}_{2^k}$ if $k$ bits are grouped to one symbol, or to arbitrary finite fields if the underlying cover domain $\mathcal{X}$ and embedding operations support $q$-ary symbols.

### 2.8.2.3 Summary

The gist of the sections on maximising embedding efficiency for the remainder of this book is twofold:

1. The actual gross message length may exceed twice the number of embedding changes.
2. For secret-key steganography[37] with sufficiently large $n$ and ratio of secure embedding positions, appropriate codes exist to concentrate the embedding changes in arbitrary locations of $x^{(0)}$ without the need to share knowledge about the embedding positions with the recipient.

Further details on coding in steganography are beyond the scope of this work.

## 2.9 Specific Detection Techniques

Up to now, contemporary techniques for digital steganography have been surveyed quite comprehensively. The remainder of this chapter is devoted to a description of the state of the art in steganalysis. This section introduces three basic techniques that have been developed specifically for the construction of steganalysis methods. Later, in Sect. 2.10, we present in greater detail a number of targeted detectors for LSB replacement steganography which are relevant to Part II of this book.

### 2.9.1 Calibration of JPEG Histograms

Calibration of JPEG histograms is a technique specific to steganalysis that was first introduced by Fridrich, Goljan, and Hogea [78] in their targeted detector against the F5 algorithm. It soon became a standard building block for many subsequent detectors against JPEG steganography, and is probably not limited to the JPEG domain, although applications in other transformed domains are rare due to the dominance of JPEG as a cover format in steganalysis research.

The idea of *calibration* is to estimate marginal statistics (histograms, co-occurrence matrices) of the *cover*'s transformed domain coefficients from the *stego* object by desynchronising the block transform structure in the spatial domain. The procedure works as depicted in Fig. 2.17. The suspected stego object in transformed domain representation is transferred back to the spatial domain (in the case of JPEG, a standard decompression operation), and then the resulting spatial domain representation is cropped by a small number

---

[37] The case for public-key steganography is less well understood, as pointed out in [16].

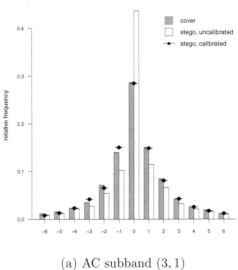

(a) AC subband $(3, 1)$

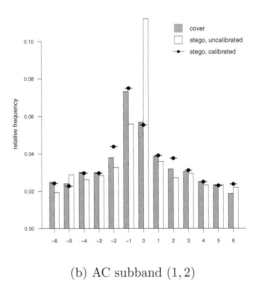

(b) AC subband $(1, 2)$

Fig. 2.16: Histograms of selected DCT subbands for a single JPEG image
$(q = 0.8)$. Its stego version is made by the F5 embedding operation $(p = 1)$

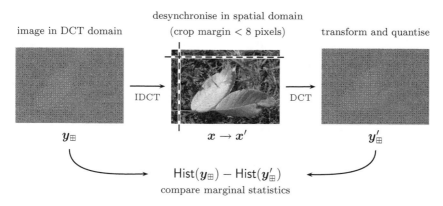

Fig. 2.17: Diagram of calibration procedure to estimate cover statistics

of pixels at two orthogonal margins. This ensures that the original ($8 \times 8$) grid is desynchronised in a subsequent transformation to the transformed domain (re-compression for JPEG, using the same quantisation matrix as before). After this sequence of operations, the coefficients exhibit marginal statistics that are much closer to the original than those of the (suspected) stego object, where the repeated application of the embedding operation might have deformed the marginal statistics.

The capability of calibration to recover original histograms is shown in Fig. 2.16 (a) for selected subbands. As expected, the stego histogram is much more leptokurtic (the frequency of 0s increases) than the cover, which is a result of the moderated-sign embedding operation of the F5 algorithm used to produce the curves (cf. Fig. 2.13 (e), p. 44). The calibration procedure recovers the original values very accurately, so evaluating the difference between uncalibrated and calibrated histograms constitutes a (crude) detector.

Interestingly, the estimation is still acceptable—albeit not perfect—for 'abnormal' (more precisely, nonzero mode) histograms, as shown in Fig. 2.16 (b). A summary measure of the calibration procedure's performance can be computed from the global histogram mean absolute error (MAE) by aggregating the discrepancy between cover and stego estimates of all 63 AC DCT subbands. Quantitative results for a set of 100 randomly selected images are reported in Fig. 2.18 for different compression qualities and margin widths. Calibrated versions of the stego objects were evaluated for crop margins between one and six pixels. The curves show the margins that led to the best (solid line) and worst (dashed line) results. Tinkering with the margin width seems to yield small but systematic improvements for high compression qualities.

These and other experimental results confirm the effectiveness and robustness of calibrating JPEG histograms, but we are not aware of a rigourous

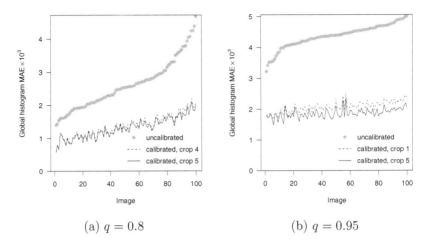

(a) $q = 0.8$                              (b) $q = 0.95$

Fig. 2.18: Mean absolute error between normalised global AC DCT coefficient histogram of 100 JPEG cover images and simulated F5 stego objects ($p = 1$) with and without calibration for two different JPEG qualities $q$. Images are sorted by increasing uncalibrated MAE

mathematical analysis of the way calibration works. Known limitations of calibration include double-compressed JPEG images (with different quantisation matrices) and images that contain *spatial resonance*. This occurs when the content has a periodicity close to (an integer multiple of) the block size of the transformation. These phenomena as well as possible remedies are discussed in [77].

### 2.9.2 Universal Detectors

Steganalysis methods can be broadly divided into *targeted detectors*, which are designed to evaluate artefacts of particular embedding operations, and *universal detectors*, which do not assume prior knowledge about a particular steganographic system. Without a specific embedding operation to reverse engineer, universal methods extract from suspected stego objects a broad set of general statistical measures (so-called *features* $\boldsymbol{f} = (f_1, \ldots, f_k)$), which likely change after embedding. Often, features from different domains (spatial, various transforms) are combined in a feature vector. Then, a classifier is trained with features from a large number of typical cover objects,[38] and classes are defined to distinguish between clean covers and stego objects. Training a classifier with a representative set of image data yields parameters $\theta$, which are then used in a second stage to assign unknown objects to classes (cover or stego objects) according to their features. Proposals for the construction of classifiers are abundant in the machine learning literature. The most important types of classifiers employed in steganalysis include

- *ordinary least-squares regression* (OLS) and its refinement for classification purposes as *Fisher linear discriminant analysis* (FLD) [59], *quadratic discriminant analysis* (QDA) [201] and generalisations to *support vector machines* (SVM) [32] for continuous features,
- *Bayesian belief networks* (BBNs) [182] for discrete or discretised features, and
- *naïve Bayes classifiers* (NBCs) [49] for mixed feature vectors.

Researchers in the area of steganalysis have combined these machine learning techniques with a variety of features extracted from different domains of images and audio files [179].

Although suffering from lower detection reliability than decent targeted detectors, universal detectors have the advantage of easy adaptability to new embedding functions. While in this case targeted detectors have to be altered or redesigned, universal detectors just require a new training. Some critics argue that universal detectors are merely a combination of features known from published targeted detectors and hence are not as 'blind' as claimed.[39] So their ability to detect fundamentally new classes of embedding functions might be limited. Although there are few breakthroughs in the development of new embedding operations, experience with the introduction of new embedding domains, such as the MP domain proposed by Cancelli et al. [36], has shown that universal detectors that did not anticipate these innovations were not able to detect this new kind of steganography reliably (see also [191] for the difficulty of detecting 'minus-F5').

Table 2.2 (p. 62) summarises a literature review of the most relevant feature sets proposed for universal detectors of image steganography in the past couple of years. Note that we omit judgements about their performance as the authors did not use comparable image sets, embedding parameters, or evaluation procedures (e.g., testing different embedding functions independently

---

[38] The training objects comprise both clean covers and stego objects generated at the design stage of the method for training purposes. This implies that developers of universal detectors typically have access to actual steganographic systems or know their embedding operations.

[39] The name *blind detector* is used synonymously for universal detectors in the literature. We prefer the term 'universal' as rarely any detector in the literature has been designed without knowledge of the (set of) target embedding operations. What is more, in digital watermarking and multimedia forensics, the term 'blind' is reserved for detectors that work without knowledge of the original cover. In this sense, targeted detectors are also blind by definition.

Table 2.2: Overview of selected universal detectors for image steganography

| Ref. feature description | classifier | # features | # images | tested stego systems |
|---|---|---|---|---|
| Avcibaş et al. [7] | | | | |
| spatial domain and spectral quality metrics | OLS | 26 | 20 | three watermarking algorithms |
| Lyu and Farid [163] | | | | |
| moments of DFT subband coefficients and size of predictor error | FLD, SVM | 72 | 1,800 | LSB, EzStego, JSteg, OutGuess |
| Harmsen and Pearlman [97] | | | | |
| HCF centre of mass (COM) | NBC | 3 | 24 | ±1, SSIS, additive noise in DCT domain for RGB images |
| Chen et al. [40] | | | | |
| DCT moments, HCF moments, DWT HCF moments of image and prediction residual | SVM | 260 | 798 | LSB, ±1, SSIS, QIM, OutGuess, F5, MB1 |
| Fridrich [68] | | | | |
| Delta to calibrated versions of DCT histogram measures, blockiness, coefficient co-occurrence | FLD | 23 | 1,814 | OutGuess, F5, MB1, MB2 |
| Goljan et al. [91] | | | | |
| higher-order moments of residual from wavelet denoising filter | FLD | 27 | 2,375 | ±1 and variants (side information, ternary codes, adaptive) |
| Shi et al. [215] | | | | |
| intra-block difference histograms of absolute DCT coefficients | SVM | 324 | 7,560 | OutGuess, F5, MB1 |
| Pevný and Fridrich [187] | | | | |
| combination of [68] and [215] | SVM | 274 | 3,400 | Jphide, Steghide, F5, OutGuess, MB1, MB2 |
| Lyu and Farid [164] | | | | |
| [163] plus LAHD phase statistics | SVM | 432 | 40,000 | JSteg, F5, Jphide, Steghide, OutGuess |
| Barbier et al. [10] | | | | |
| moments of residual entropy in Huffman-encoded blocks, KLD to reference p.d.f. | FLD | 7+ | 4,000 | F5, Jphide, OutGuess |

or jointly). Another problem is the risk of overfitting when the number of images in the training and test set is small compared to the number of features, and all images are taken from a single source. In these cases, the parameters of the trained classifier are estimated with high standard errors and may be adapted too much to the characteristics of the test images so that the results do not generalise.

Although machine learning techniques were first used in steganalysis to construct universal detectors, they become increasingly common as tools for constructing targeted detectors as well. This is largely for convenience reasons: if several metrics sensitive to embedding are identified, but their optimal combination is unknown, then machine learning techniques help to find good decision rules quickly (though they are sometimes hard to explain). The $\pm 1$ detector proposed by Boncelet and Marvel [28] and the targeted detector of MB2 by Ullerich [227] are representatives of this approach.

The research in this book is restricted to targeted detectors, mainly because they have better performance than universal detectors and their higher transparency facilitates reasoning about dependencies between cover properties and detection performance.

## 2.9.3 Quantitative Steganalysis

The attribute *quantitative* in steganalysis means that the detector outputs not only a binary decision, but an estimate of the lengths of the secret message, which can be zero for clean covers [79]. This implies that those methods are still reliable when only parts of the cover's steganographic capacity have been used (early statistical detectors could only detect reliably messages with full capacity or imperfect spreading [238]).

We define quantitative detectors as functions that estimate the net embedding rate $p$. The attribute 'net' means that possible gains in embedding efficiency due to message coding (see Sect. 2.8.2) are not taken into account,

$$\hat{p} = \mathsf{Detect_{Quant}}(\boldsymbol{x}^{(p)}). \tag{2.24}$$

A useful property of quantitative detectors is that detection performance can be measured more granularly than mere error rates, e.g., by comparing the estimated embedding rate $p$ with the estimate $\hat{p}$. Quantitative detectors for a particular embedding operation, namely LSB replacement, play an important role in the specific results presented in Part II. Therefore, we introduce three state-of-the-art detectors and some variants in the next section.

## 2.10 Selected Estimators for LSB Replacement in Spatial Domain Images

We follow the terminology of Ker [120] and call a quantitative detector *estimator* when we refer to its ability to determine the secret message length, and *discriminator* when we focus on separating stego from cover objects.

### 2.10.1 RS Analysis

RS analysis,[40], developed by Fridrich, Goljan, and Du [74], estimates the number of embedding changes by measuring the proportion of *regular* and *singular* non-overlapping $k$-tuples (groups) of spatially adjacent pixels before and after applying three types of flipping operations:

1. $\mathsf{Flip}^{+1} : \mathcal{X} \to \mathcal{X}$ is a bijective mapping between pairs of values that mimics exactly the embedding operation of LSB replacement: $0 \leftrightarrow 1, 2 \leftrightarrow 3, \ldots$
2. $\mathsf{Flip}^{-1} : \mathcal{X} \to \mathcal{X}$ is a bijective mapping between the opposite (shifted) pairs, that is, $\mathsf{Flip}^{-1}(x) = \mathsf{Flip}^{+1}(x+1) - 1$; hence, $-1 \leftrightarrow 0, 1 \leftrightarrow 2, \ldots$
3. $\mathsf{Flip}^{0} : \mathcal{X} \to \mathcal{X}$ is the identity function.

Groups are counted as *regular* and assigned to multi-set $\mathcal{R}_{\mathbf{m}}$ if the value of a discrimination function $\mathsf{Discr} : \mathcal{X}^k \to \mathbb{R}$ increases after applying $\mathsf{Flip}^{m_i}$ on the individual pixels of the group according to a mask vector $\mathbf{m} \in \{0,1\}^k$, i.e.,

$$\mathsf{Discr}(\boldsymbol{x}) < \mathsf{Discr}\left(\mathsf{Flip}^{m_1}(x_1), \mathsf{Flip}^{m_2}(x_2), \ldots, \mathsf{Flip}^{m_k}(x_k)\right). \tag{2.25}$$

Conversely, multi-set $\mathcal{S}_{\mathbf{m}}$ contains all so-called *singular* groups, by definition, when

$$\mathsf{Discr}(\boldsymbol{x}) > \mathsf{Discr}\left(\mathsf{Flip}^{m_1}(x_1), \mathsf{Flip}^{m_2}(x_2), \ldots, \mathsf{Flip}^{m_k}(x_k)\right). \tag{2.26}$$

The remaining *unusable* groups, for which none of inequalities (2.25) and (2.26) hold, is disregarded in the further analysis. The suggested implementation for the discrimination function is a noisiness measure based on the $L_1$-norm, but other summary functions are possible as well:

$$\mathsf{Discr}(\boldsymbol{u}) = \sum_{i=2}^{|\boldsymbol{u}|} |u_i - u_{i-1}|. \tag{2.27}$$

Figure 2.19 shows the typical shape of the relative sizes of $\mathcal{R}_{\mathbf{m}}$ (solid black curve) and $\mathcal{S}_{\mathbf{m}}$ (solid grey curve) as a function of the fraction of flipped LSBs

---

[40] RS stands for *regular/singular* named after the concept of regular and singular groups of pixels.

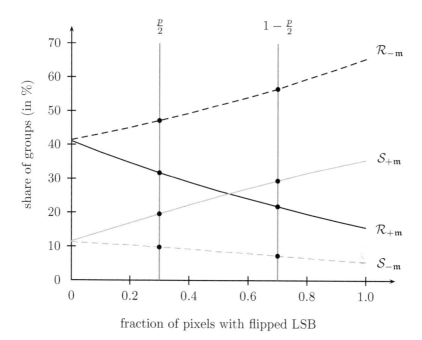

Fig. 2.19: Typical RS diagram of a single image: relative size of sets of regular ($\mathcal{R}$) and singular ($\mathcal{S}$) groups for direct ($+\mathbf{m}$) and inverse ($-\mathbf{m}$) mask $\mathbf{m} = (0, 1, 1, 0)$ as a function of the fraction of flipped LSBs

for a single image with non-overlapping horizontal groups of size $k = 4$ and mask $\mathbf{m} = (0, 1, 1, 0)$. The corresponding dashed curves $\mathcal{R}_{-\mathbf{m}}$ and $\mathcal{S}_{-\mathbf{m}}$ result from applying the *inverse mask* $-\mathbf{m} = (0, -1, -1, 0)$. LSB replacement is detectable because the proportion of regular and singular groups deviates in the opposite direction with increasing number of flipped LSBs.

The unknown embedding rate $p$ of a suspect image $\boldsymbol{x}^{(p)}$ can be estimated from observable quantities in this diagram, a linear approximation of the 'outer' $\mathcal{R}_{-\mathbf{m}}$ and $\mathcal{S}_{-\mathbf{m}}$ curves as well as a quadratic approximation of the 'inner' curves $\mathcal{R}_{+\mathbf{m}}$ and $\mathcal{S}_{+\mathbf{m}}$.[41] The net embedding rate $\hat{p}$ is approximately half of the fraction of pixels with flipped LSBs.[42]

- The size of $\mathcal{R}_{+\mathbf{m}}$, $\mathcal{R}_{-\mathbf{m}}$, $\mathcal{S}_{+\mathbf{m}}$ and $\mathcal{S}_{-\mathbf{m}}$ at the intersection with the vertical line $p/2$ can be obtained directly from $\boldsymbol{x}^{(p)}$.

---

[41] The linear and quadratic shapes of the curves has been proven for groups of size $k = 2$ in [50]. More theory on the relation between the degree of the polynomial and the group size $k$ is outlined in the introduction of [120].

[42] Net embedding rate and secret message length as a fraction of cover size $n$ differ if efficiency-enhancing coding is employed; see Sect. 2.9.3.

- Flipping the LSBs of *all* samples in $\boldsymbol{x}^{(p)}$ and the subsequent calculation of multi-set sizes yield an indirect measure of the sizes of $\mathcal{R}_{+\mathbf{m}}$, $\mathcal{R}_{-\mathbf{m}}$, $\mathcal{S}_{+\mathbf{m}}$ and $\mathcal{S}_{-\mathbf{m}}$ at the intersection with the vertical line $1 - p/2$.

Further, two assumptions,

1. the two pairs of curves $\mathcal{R}_{\pm\mathbf{m}}$ and $\mathcal{S}_{\pm\mathbf{m}}$ intersect at 0 (a plausible assumption if we reckon that the distribution of intensity values in the image acquisition process is invariant to small additive constants), and
2. curves $\mathcal{R}_{+\mathbf{m}}$ and $\mathcal{S}_{+\mathbf{m}}$ intersect at 50% flipped LSBs (justified in [74] and [79] with a theorem cited from [90] saying that "the lossless capacity in the LSBs of a fully embedded image is zero"; in practice, this assumption is violated more frequently than the first one),

are sufficient to find a unique[43] solution for $\hat{p} = \frac{z}{z - 1/2}$.

Auxiliary variable $z$ is the smaller root of the quadratic equation

$$2(\Delta_{+\mathbf{m}} + \Delta'_{+\mathbf{m}})z^2 + (\Delta'_{-\mathbf{m}} - \Delta_{-\mathbf{m}} - \Delta_{+\mathbf{m}} - 3\Delta'_{+\mathbf{m}})z - \Delta'_{-\mathbf{m}} + \Delta'_{+\mathbf{m}} = 0 \quad (2.28)$$

with    $\Delta_{\mathbf{m}} = \dfrac{k}{n} \cdot (|\mathcal{R}_{\mathbf{m}}| - |\mathcal{S}_{\mathbf{m}}|)$ at $\dfrac{p}{2}$    (computed from $\boldsymbol{x}^{(p)}$), and

$\Delta'_{\mathbf{m}} = \dfrac{k}{n} \cdot (|\mathcal{R}_{\mathbf{m}}| - |\mathcal{S}_{\mathbf{m}}|)$ at $1 - \dfrac{p}{2}$    (computed from $\mathsf{Flip}^{+1}(\boldsymbol{x}^{(p)})$).

For $p$ close to 1, cases where Eq. (2.28) has no real root occur more frequently. In such cases we set $\hat{p} = 1$ because the suspect image is almost certainly a stego image. However, failures of the RS estimation equation have to be borne in mind when evaluating the distribution of RS estimates and estimation errors $\hat{p} - p$, as done in Chapter 5.

The way pixels are grouped (topology and overlap), group size $k$, mask vector $\mathbf{m}$ and the choice of the discrimination function $\mathsf{Discr}$ (Eq. 2.27) are subject to experimental fine tuning. Empirical results can be found in [118] and [119]. Note that global RS estimates are not reliable if the message is not distributed randomly in the stego image. In this case a moving window variant of RS or SPA, as suggested in [79], or more efficient sequential variants of WS analysis [128, 133] are preferable.

---

[43] Yet another set of quantities could be obtained for 50% flipped LSBs by averaging over repeated randomisations of the entire LSB plane. Incorporating this information leads to an over-specified equation system for which a least-squares solution can be found to increase the robustness against measurement errors of individual quantities. Alternatively, the zero-intersection assumption can be weakened. Although there is little documented evidence on whether the performance gains justify the additional effort, the dominant variant of RS follows the approach described above. Research on RS improvements has stalled since more reliable detectors for LSB replacement have been invented.

## 2.10.2 Sample Pair Analysis

The steganalysis method known as *sample pair analysis*[44] (SPA) was first introduced by Dumitrescu et al. [50, 51]. In our presentation of the method we adapt the more extensible alternative notation of Ker [120] to our conventions.[45]

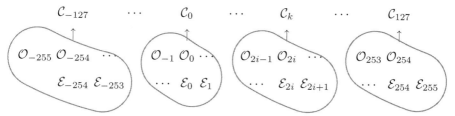

Fig. 2.20: Relation of trace sets and subsets in SPA ($\mathcal{X} = [0, 255]$)

Similarly to RS analysis, SPA evaluates groups of spatially adjacent pixels. It assigns each pair $(x_1, x_2)$ to a *trace set* $\mathcal{C}_i$, so that

$$\mathcal{C}_i = \left\{ (x_1, x_2) \in \mathcal{X}^2 \; \middle| \; \left\lfloor \frac{x_2}{2} \right\rfloor - \left\lfloor \frac{x_1}{2} \right\rfloor = i \right\}, \quad |i| \leq \lfloor (\max \mathcal{X} - \min \mathcal{X})/2 \rfloor.$$
$$(2.29)$$

Each trace set $\mathcal{C}_i$ can be further partitioned into up to four *trace subsets*, of which two types can be distinguished:

- Pairs $(x_1, x_2)$ whose values differ by $i = x_2 - x_1$ and whose first elements $x_1$ are *even* belong to $\mathcal{E}_i$.
- Pairs $(x_1, x_2)$ whose values differ by $i = x_2 - x_1$ and whose first elements $x_1$ are *odd* belong to $\mathcal{O}_i$.

Consequently, the union of trace subsets $\mathcal{E}_{2i+1} \cup \mathcal{E}_{2i} \cup \mathcal{O}_{2i} \cup \mathcal{O}_{2i-1} = \mathcal{C}_i$ constitutes a trace set (cf. Fig. 2.20). This definition of trace sets and subsets ensures that the LSB replacement embedding operation never changes a sample pair's trace set, i.e., $\mathcal{C}_i^{(0)} = \mathcal{C}_i^{(p)} = \mathcal{C}_i$, but may move sample pairs between trace subsets that constitute the same trace set. So cardinalities $|\mathcal{C}_i|$ are invariant to LSB replacement, whereas $|\mathcal{E}_i|$ and $|\mathcal{O}_i|$ are sensitive. The transition probabilities between trace subsets depend on the net embedding rate $p$ as depicted in the transition diagram of Fig. 2.21. So, the effect of

---

[44] The same method is sometimes also referred to as *couples analysis* in the literature to avoid possible confusion with *pairs analysis* by Fridrich et al. [82], another method not relevant in this book. Therefore, we stick to the original name.

[45] This presentation minds the order of samples in each pair; hence, $i$ can be negative. The original publication made no difference between pairs $(u, v)$ and $(v, u)$. This led to a special case for $\lfloor u/2 \rfloor = \lfloor v/2 \rfloor$.

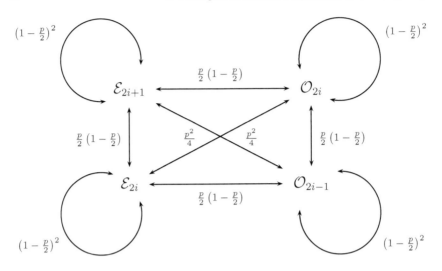

Fig. 2.21: Transition diagram between trace subsets under LSB replacement

applying LSB replacement with rate $p$ on the expected cardinalities of the trace subsets can be written as four quadratic equations (in matrix notation):

$$
\begin{bmatrix} |\mathcal{E}^{(p)}_{2i+1}| \\ |\mathcal{E}^{(p)}_{2i}| \\ |\mathcal{O}^{(p)}_{2i}| \\ |\mathcal{O}^{(p)}_{2i-1}| \end{bmatrix}
=
\begin{bmatrix}
\left(1-\frac{p}{2}\right)^2 & \frac{p}{2}\left(1-\frac{p}{2}\right) & \frac{p}{2}\left(1-\frac{p}{2}\right) & \frac{p^2}{4} \\
\frac{p}{2}\left(1-\frac{p}{2}\right) & \left(1-\frac{p}{2}\right)^2 & \frac{p^2}{4} & \frac{p}{2}\left(1-\frac{p}{2}\right) \\
\frac{p}{2}\left(1-\frac{p}{2}\right) & \frac{p^2}{4} & \left(1-\frac{p}{2}\right)^2 & \frac{p}{2}\left(1-\frac{p}{2}\right) \\
\frac{p^2}{4} & \frac{p}{2}\left(1-\frac{p}{2}\right) & \frac{p}{2}\left(1-\frac{p}{2}\right) & \left(1-\frac{p}{2}\right)^2
\end{bmatrix}
\begin{bmatrix} |\mathcal{E}^{(0)}_{2i+1}| \\ |\mathcal{E}^{(0)}_{2i}| \\ |\mathcal{O}^{(0)}_{2i}| \\ |\mathcal{O}^{(0)}_{2i-1}| \end{bmatrix}.
$$
$$(2.30)$$

Trace subsets $\mathcal{E}^{(p)}$ and $\mathcal{O}^{(p)}$ are observable from a given stego object. An approximation of the cardinalities of the cover trace subsets $\mathcal{E}^{(0)}$ and $\mathcal{O}^{(0)}$ can be rearranged as a function of $p$ by inverting Eq. (2.30). The transition matrix is invertible for $p < 1$:

$$
\begin{bmatrix} |\hat{\mathcal{E}}^{(0)}_{2i+1}| \\ |\hat{\mathcal{E}}^{(0)}_{2i}| \\ |\hat{\mathcal{O}}^{(0)}_{2i}| \\ |\hat{\mathcal{O}}^{(0)}_{2i-1}| \end{bmatrix}
=
\frac{1}{(2-2p)^2}
\begin{bmatrix}
(2-p)^2 & p(p-2) & p(p-2) & p^2 \\
p(p-2) & (2-p)^2 & p^2 & p(p-2) \\
p(p-2) & p^2 & (2-p)^2 & p(p-2) \\
p^2 & p(p-2) & p(p-2) & (2-p)^2
\end{bmatrix}
\begin{bmatrix} |\mathcal{E}^{(p)}_{2i+1}| \\ |\mathcal{E}^{(p)}_{2i}| \\ |\mathcal{O}^{(p)}_{2i}| \\ |\mathcal{O}^{(p)}_{2i-1}| \end{bmatrix}.
$$
$$(2.31)$$

With one additional cover assumption, namely $|\mathcal{E}^{(0)}_{2i+1}| \approx |\mathcal{O}^{(0)}_{2i+1}|$, the first equation of this system for $i$ can be combined with the fourth equation for $i+1$ to obtain a quadratic estimator $\hat{p}$ for $p$. This assumption mirrors the first assumption of RS analysis (see p. 66). It is plausible because cardinalities of sample pairs in natural images should not depend on the parity of their first element:

$$|\hat{\mathcal{E}}_{2i+1}^{(0)}| = |\hat{\mathcal{O}}_{2i+1}^{(0)}| \tag{2.32}$$

$$0 = \frac{(2-p)^2}{(2-2p)^2}\left(|\mathcal{E}_{2i+1}^{(p)}| - |\mathcal{O}_{2i+1}^{(p)}|\right) + \frac{p^2}{(2-2p)^2}\left(|\mathcal{O}_{2i-1}^{(p)}| - |\mathcal{E}_{2i+3}^{(p)}|\right) +$$
$$\frac{p(p-2)}{(2-2p)^2}\left(|\mathcal{E}_{2i}^{(p)}| + |\mathcal{O}_{2i}^{(p)}| - |\mathcal{E}_{2i+2}^{(p)}| - |\mathcal{O}_{2i+2}^{(p)}|\right) \tag{2.33}$$

$$0 = p^2\left(|\mathcal{C}_i| - |\mathcal{C}_{i+1}|\right) + 4\left(|\mathcal{E}_{2i+1}^{(p)}| - |\mathcal{O}_{2i+1}^{(p)}|\right) +$$
$$2p\left(|\mathcal{E}_{2i+2}^{(p)}| + |\mathcal{O}_{2i+2}^{(p)}| - 2|\mathcal{E}_{2i+1}^{(p)}| + 2|\mathcal{O}_{2i+1}^{(p)}| - |\mathcal{E}_{2i}^{(p)}| - |\mathcal{O}_{2i}^{(p)}|\right) \tag{2.34}$$

The smaller root of Eq. (2.34) is a secret message length estimate $\hat{p}_i$ based on the information of pairs in trace set $\mathcal{C}_i$. Standard SPA sums up the family of estimation equations (2.34) for a fixed interval around $\mathcal{C}_0$, such as $-30 \leq i \leq 30$, and calculates a single root $\hat{p}$ from the aggregated quadratic coefficients. Experimental results from fairly general test images have shown that standard SPA, using all overlapping horizontal and vertical pairs of greyscale images, is slightly more accurate than standard RS analysis [22, 118]. For solely discrimination purposes (hence, ignoring the quantitative capability), it has been found that smarter combinations of individual roots for small $|i|$, e.g., $\hat{p}^* = \min(\hat{p}_{-2}, \ldots, \hat{p}_2)$, can improve SPA's detection performance further [118].

Similarly to RS, Eq. (2.34) may fail to produce real roots, which happens more frequently as $p$ approaches 1. In these cases, the tested object is almost certainly a stego image, but the exact message length cannot be determined.

### 2.10.3 Higher-Order Structural Steganalysis

Sample pair analysis, as presented in Sect. 2.10.2, is a specific representative of a family of detectors for LSB replacement which belong to the general framework of *structural steganalysis*. The attribute 'structural' refers to the design of detectors to deliberately exploit, at least in theory, all combinatorial measures of the artificial dependence between sample differences and the parity structure that is typical for LSB replacement.[46] A common element in all structural detectors is to estimate $\hat{p}$ so that macroscopic cover properties,

---

[46] Under LSB replacement (see Eq. 2.8), even cover samples are never decremented whereas odd cover samples are never incremented. This leads to the artificial parity structure.

which can be approximated from the stego object by inverting the effects of embedding as a function of $p$, match cover assumptions best. Hence, also RS analysis and the method by Zhang and Ping [252] (disregarded in this book) can be subsumed as (less canonical) representatives of the structural framework.[47] In this section we review three important alternative detectors of the structural framework, which are all presented as extensions to SPA.

### 2.10.3.1 Least-Squares Solutions to SPA

The combination of individual equations (2.34) for different $i$, as suggested in the original publication [51], appears a bit arbitrary. Lu et al. [160] have suggested an alternative way to impose the cover assumption $|\mathcal{E}_{2i+1}| \approx |\mathcal{O}_{2i+1}|$. Instead of setting both cardinalities equal, they argue that the difference between odd and even trace subsets should be interpreted as error,

$$\epsilon_i = |\mathcal{E}_{2i+1}| - |\mathcal{O}_{2i+1}|, \tag{2.35}$$

and a more robust estimate for $\hat{p}$ can be found by minimising the squared errors $\hat{p} = \arg\min_p \sum_i \epsilon_i^2$, which turns out to be a solution to a cubic equation. Note that the least-squares method (LSM) implicitly attaches a higher weight to larger trace subsets (those with small $|k|$ in natural images), where higher absolute deviations from the cover assumption are observable. Quantitative results reported in [160] confirm a higher detection accuracy in terms of MAE and estimator standard deviation than both RS and standard SPA for three image sets throughout all embedding rates $p$. In practice, pure LSM has shown to cause severe inaccuracies when $p$ is close to 1, so a combination with standard SPA to screen for large embedding rates by a preliminary estimate is recommended in [22]. The combined method is called SPA/LSM.

### 2.10.3.2 Maximum-Likelihood Solutions to SPA

The process an image undergoes from acquisition via embedding to a stego object is indeterministic at many stages. The choice of the embedding positions and the encrypted message bits are (pseudo)random by definition to achieve secrecy. Additional parameters unknown to the steganalyst have to be modelled as random variables as well, foremost the cover realisation and the actual embedding rate $p$. A common simplification in the construction of

---

[47] At the time of this writing, it is unclear whether WS analysis (to be presented in the following section) belongs to the structural class (it probably does). WS was not well recognised when the structural terminology was introduced, so it is not commented on in [120]. Its different approach justifies it being treated as something special. However, variants of WS can be found that have a striking similarity to RS or SPA.

structural detectors is the (implicit) reduction of random variables to expectations. This is suboptimal as it ignores the shape of the random variables' probability functions, and their ad hoc algebraic combination may deviate from the true joint distribution. Moreover, deviations from the expectation are not weighted by the size of the standard error, which differs as trace sets are sparser populated for large $|i|$. As a remedy, Ker [126] has replaced the cover assumption $|\mathcal{E}_{2i+1}| = |\mathcal{O}_{2i+1}|$ by a probabilistic model in which all pairs in the union set $\mathcal{D}_{2i+1} = \mathcal{E}_{2i+1} \cup \mathcal{O}_{2i+1}$ are distributed uniformly into subsets $\mathcal{E}_{2i+1}$ and $\mathcal{O}_{2i+1}$ during an imaginary image acquisition process. The term 'pre-cover' has been suggested for the imaginary low-precision image composed of pairs in $\mathcal{D}_i$. With this model, probability functions for all random variables can be defined under gentle assumption and thus a likelihood function for structural detectors can be derived. Estimating $\hat{p}$ reduces to maximising the likelihood (ML).[48] As an additional advantage, likelihood ratio tests (LRTs) allow mathematically well-founded hypothesis tests for the existence of a stego message $p > 0$ against the null hypothesis $p = 0$ (though no practical tests exist that perform better than discriminators by the estimate $\hat{p}$, yet [126]).

Performance evaluations of a single implementation of SPA/ML suggest that ML estimates are much more accurate than other structural detectors, especially for low embedding rates $p$, where accuracy matters for discriminating stego images from plain covers. Unfortunately, the numerical complexity of ML estimates is high due to a large number of unknown parameters and the intractability of derivatives with respect to $p$. Computing a single SPA/ML estimate of a 1.0 megapixel image takes about 50 times longer than a standard SPA estimate [126]. However, more efficient estimation strategies using iteratively refined estimates for the unknown cardinalities $|\mathcal{D}_i|$ (e.g., via the expectation maximisation algorithm [47]) are largely unexplored and promise efficiency improvements in future ML-based methods. All in all, structural ML estimators are rather novel and leave open questions for research.

Earlier non-structural proposals for maximum-likelihood approaches to detect LSB replacement in the spatial domain [46, 48] work solely on the first and second order (joint) histograms and are less reliable than the ML-variant of SPA, which uses trace subsets to exploit the characteristic parity structure.

### 2.10.3.3 Triples and Quadruples Analysis

The class of structural detectors can be extended by generalising the principles of SPA from pairs to $k$-tuples [120, 122]. Hence, trace sets and subsets are indexed by $k - 1$ suffixes and the membership rules generalise as follows:

---

[48] As argued in [126], the least-squares solution concurs with the ML estimate only in the case of independent Gaussian variables, but the covariance matrix contains nonzero elements for structural detectors.

$$\mathcal{C}_{i_1,\ldots,i_{k-1}} = \left\{ (x_1,\ldots,x_k) \in \mathcal{X}^k \;\middle|\; \left\lfloor \frac{x_{j+1}}{2} \right\rfloor - \left\lfloor \frac{x_j}{2} \right\rfloor = i \; \forall j : 1 \le j < k \right\}$$

$$\mathcal{E}_{i_1,\ldots,i_{k-1}} = \left\{ (x_1,\ldots,x_k) \in \mathcal{X}^k \;\middle|\; x_{j+1} - x_j = i \; \forall j : 1 \le j < k \; \wedge x_1 \text{ even} \right\}$$

$$\mathcal{O}_{i_1,\ldots,i_{k-1}} = \left\{ (x_1,\ldots,x_k) \in \mathcal{X}^k \;\middle|\; x_{j+1} - x_j = i \; \forall j : 1 \le j < k \; \wedge x_1 \text{ odd} \right\}$$

Each trace set contains $2^k$ trace subsets. The generalisation of the transition matrix of Eq. (2.30) is given by the iterative rule $\boldsymbol{t}_k(p) = \boldsymbol{t}_{k-1}(p) \otimes \boldsymbol{t}_1(p)$ with initial condition

$$\boldsymbol{t}_1 = \begin{bmatrix} 1 - \frac{p}{2} & \frac{p}{2} \\ \frac{p}{2} & 1 - \frac{p}{2} \end{bmatrix}. \tag{2.36}$$

For example, when $k = 3$, each trace set is divided into eight trace subsets with transition probabilities

- $\left(1 - \frac{p}{2}\right)^3$ for remaining in the same trace subset (no LSB flipped),
- $\frac{p}{2}\left(1 - \frac{p}{2}\right)^2$ for a move into a subset that corresponds to a single LSB flip,
- $\frac{p^2}{4}\left(1 - \frac{p}{2}\right)$ for a move into a subset where two out of three LSBs are flipped, and
- $\frac{p^3}{8}$ for a move to the 'opposite' trace subsets, i.e., with all LSBs flipped.

The corresponding transition diagram is depicted in Fig. 2.22. Selected transition paths are plotted and annotated only for trace subset $\mathcal{O}_{2i-1,2j}$ to keep the figure legible.

Inverting the transition matrix is easy following the procedure of [120]. A more difficult task for higher-order structural steganalysis is finding (all) equivalents for the cover assumption $|\mathcal{E}_{x_1,\ldots,x_{k-1}}| \approx |\mathcal{O}_{x_1,\ldots,x_{k-1}}|$. Apart from this *parity symmetry*, Ker [122] has identified two more classes of plausible cover assumptions, which he calls *inversion symmetry* and *permutative symmetry*. Once all relevant symmetries are identified, the respective estimation equations similar to Eq. (2.34) can be derived and solved either by ad hoc summation, the above-described least-squares fit, or through an ML estimate.

In general, higher-orders of structural steganalysis yield moderate performance increases, especially for low embedding rates, but for increasing $k$, their applicability reduces to even lower ranges of $p$. Another drawback of higher-orders is the low number of observations in each subset, which increasingly thwarts the use of the law of large numbers that frequencies converge towards their expected value, and the normal approximation for the multinomial distributions in the ML estimator. So, we conjecture that the optimal order $k$ should depend on the size of the stego objects under analysis.

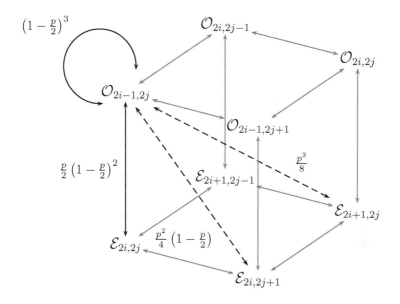

Fig. 2.22: Transition cube of trace subsets for Triples analysis ($k = 3$)

## 2.10.4 Weighted Stego Image Steganalysis

The steganalysis method using a weighted stego image (WS) proposed by Fridrich and Goljan [73] in 2004 differs from the above-discussed methods in several aspects: it is a mathematically better founded, modular, and computationally fast estimator for the net embedding rate of LSB replacement steganography in the spatial domain. In its original form, its performance is competitive with alternative methods only at high embedding rates, where high accuracy is less relevant in practice. Thus, the method resided in the shade for years. In this section we describe *standard WS* in an extensible notation. Improvements of the method are presented in Chapter 6.

WS analysis is based on the following concepts:

- A *weighted stego image* with scalar parameter $\lambda$:

$$x^{(p,\lambda)} = \lambda \overline{x}^{(p)} + (1 - \lambda) x^{(p)}, \qquad (2.37)$$

where $\overline{x} = x + (-1)^x = \mathsf{Flip}^{+1}(x)$, also applicable to vectors $x$, is defined as a sample with inverted LSB to simplify the notation.

- Function $\mathsf{Pred} : \mathcal{X}^n \to \mathcal{X}^n$, a local predictor for pixels in cover images from their spatial neighbourhood.

- Function $\mathsf{Conf} : \mathcal{X}^n \to \mathbb{R}^{+^n}$, a measure of local predictability with respect to $\mathsf{Pred}$. By convention, lower values denote higher confidence or predictability.

The WS method is modular as $\mathsf{Pred}$ and $\mathsf{Conf}$ can be adapted to specific cover models while maintaining the same underlying logic of the estimator. Theorem 1 of [73] states the key idea of WS, namely that $\hat{p}$ can be estimated via the weight $\lambda$ that minimises the Euclidean distance between the weighted stego image $\boldsymbol{x}^{(p,\lambda)}$ and the cover $\boldsymbol{x}^{(0)}$:

$$\hat{p} = 2 \arg\min_{\lambda} \sum_{i=1}^{n} \left( x^{(p,\lambda)} - x^{(0)} \right)^2. \tag{2.38}$$

The proof of this theorem is repeated in Appendix C, using our notation. In practice, the steganalyst does not know the cover $\boldsymbol{x}^{(0)}$, so it has to be estimated from the stego object $\boldsymbol{x}^{(p)}$ itself. According to Theorem 3 in [73], the relation in Eq. (2.38) between $\hat{p}$ and $\lambda$ still holds approximately if

1. $\boldsymbol{x}^{(0)}$ is replaced by its prediction $\mathsf{Pred}(\boldsymbol{x}^{(p)})$, and (independently)
2. the $L_2$-norm itself is weighted by vector $\boldsymbol{w}$ to reflect heterogeneity in predictability of individual samples.[49]

So, we obtain the main estimation equation that is common to all WS methods:

$$\hat{p} = 2 \arg\min_{\lambda} \sum_{i=1}^{n} w_i \left( x_i^{(p,\lambda)} - \mathsf{Pred}(\boldsymbol{x}^{(p)})_i \right)^2 \tag{2.39}$$

$$= 2 \arg\min_{\lambda} \sum_{i=1}^{n} w_i \left( \lambda \bar{x}_i^{(p)} + (1 - \lambda) x_i^{(p)} - \mathsf{Pred}(\boldsymbol{x}^{(p)})_i \right)^2$$

$$= 2 \sum_{i=1}^{n} w_i \left( x_i^{(p)} - \bar{x}_i^{(p)} \right) \left( x_i^{(p)} - \mathsf{Pred}(\boldsymbol{x}^{(p)})_i \right), \tag{2.40}$$

where weights $\boldsymbol{w} = (w_1, \ldots, w_n)$ are calculated from the predictability measure as follows:

$$w_i \propto \frac{1}{1 + \mathsf{Conf}(\boldsymbol{x}^{(p)})_i} \quad , \text{ so that } \quad \sum_{i=1}^{n} w_i = 1. \tag{2.41}$$

In standard WS, function $\mathsf{Pred}$ is instantiated as the unweighted mean of the four directly adjacent pixels (in horizontal and vertical directions, ignoring diagonals). More formally,

---

[49] These optional local weights $w_i$ should not be confused with the global weight $\lambda$ that lends its name to the method. This is why the seemingly counterintuitive term 'unweighted weighted stego image steganalysis' makes sense: it refers to WS with constant local weight $w_i = 1/n \; \forall i$ (still using an estimation via $\lambda$).

$$\mathsf{Pred}(\boldsymbol{x}) = \boldsymbol{\Phi}\, \boldsymbol{x} \oslash \boldsymbol{\Phi}\, \mathbf{1}_{n \times 1}, \tag{2.42}$$

where $\boldsymbol{\Phi}$ is a $n \times n$ square matrix and $\Phi_{ij} = 1$ if the sample $\boldsymbol{x}_j^{(p)}$ is an upper, lower, left or right direct neighbour of sample $\boldsymbol{x}_i^{(p)}$; otherwise, $\Phi_{ij} = 0$. Operator $\oslash$ denotes element-wise division. Consistent with the choice of $\mathsf{Pred}$, function $\mathsf{Conf}$ measures predictability as the empirical variance of all pixels in the local predictor; thus,

$$\mathsf{Conf}(\boldsymbol{x}) = \left(\frac{1}{n}\right) \left[ \left((\boldsymbol{x} \otimes \mathbf{1}_{1 \times n}) \odot \boldsymbol{\Phi}\right)^2 \mathbf{1}_{n \times 1} \right] - \left(\frac{1}{n^2}\right) \left[ \left((\boldsymbol{x} \otimes \mathbf{1}_{1 \times n}) \odot \boldsymbol{\Phi}\right) \mathbf{1}_{n \times 1} \right]^2 \tag{2.43}$$

It is important to note that both the local prediction $\mathsf{Pred}$ and the local weights $w_i$ must not depend on the value of $x_i^{(p)}$. Otherwise, correlation between the predictor error in covers $\mathsf{Pred}(\boldsymbol{x}^{(0)}) - \boldsymbol{x}^{(0)}$ and the parity of the stego sample $\boldsymbol{x}^{(p)} - \overline{\boldsymbol{x}}^{(p)}$ accumulates to a non-negligible error term in the estimation relation Eq. (2.40), which can be rewritten as follows to study the error components (cf. Eq. 6 of [73]):

$$\hat{p} = \overbrace{2 \sum_{i=1}^{n} w_i \left( x_i^{(p)} - \overline{x}_i^{(p)} \right) \left( x_i^{(p)} - x_p^{(0)} \right)}^{\approx\, p} + \tag{2.44}$$

$$2 \sum_{i=1}^{n} w_i \left( x_i^{(p)} - \overline{x}_i^{(p)} \right) \Big( \underbrace{x_p^{(0)} - \mathsf{Pred}(\boldsymbol{x}^{(0)})_i}_{\text{predictor error}} + \underbrace{\mathsf{Pred}(\boldsymbol{x}^{(0)})_i - \mathsf{Pred}(\boldsymbol{x}^{(p)})_i}_{\text{predicted stego noise}} \Big).$$

Choosing functions $\mathsf{Pred}$ and $\mathsf{Conf}$ to be independent of the centre pixel bounds the term annotated as 'predictor error'. The term 'predicted stego noise' causes an estimation *bias* in images with large connected areas of constant pixel intensities,[50] for example, as a result of saturation. Imagine a cover where all pixels are constant and even, $x_i^{(0)} = 2k\ \forall i$ with $k$ integer. With $\mathsf{Pred}$ as in Eq. (2.42), the prediction error in the cover $x_i^{(0)} - \mathsf{Pred}(\boldsymbol{x}^{(0)})_i = 0$, but the predicted stego noise $\mathsf{Pred}(\boldsymbol{x}^{(0)})_i - \mathsf{Pred}(\boldsymbol{x}^{(p)})_i$ is negative on average because $\mathsf{Pred}(\boldsymbol{x}^{(0)})_i = 2k\ \forall i$ and $\mathsf{Pred}(\boldsymbol{x}^{(p)})_i = 2k$ with probability $(1 - p/2)^4$ (none of the four neighbours flipped), or $2k < \mathsf{Pred}(\boldsymbol{x}^{(p)})_i \le 2k{+}1$ otherwise. With $w_i = 1/n\ \forall i$, the remaining error term,

$$\frac{2}{n} \sum_{i=1}^{n} \left( x_i^{(p)} - \overline{x}_i^{(p)} \right) \left( \mathsf{Pred}(\boldsymbol{x}^{(0)})_i - \mathsf{Pred}(\boldsymbol{x}^{(p)})_i \right) > 0 \quad \text{for } p > 0, \tag{2.45}$$

cancels out only for $p \in \{0, 1\}$. The size of the bias in real images depends on the proportion of flat areas relative to the total image size. Fridrich and

---

[50] Later, in Chapter 6, we argue that a more precise criterion than flat pixels is a phenomenon we call *parity co-occurrence*, which was not considered in the original publication.

Goljan [73] propose a heuristic bias correction, which estimates the number of flat pixels in $x^{(0)}$ from the number of flat pixels in $x^{(p)}$, although they acknowledge that their estimate is suboptimal as flat pixels can also appear randomly in $x^{(p)}$ if the cover pixel is not flat. While this correction apparently removes outliers in the test images of [73], we could not reproduce improvements of estimation accuracy in our own experiments.

Compared to other quantitative detectors for LSB replacement, WS estimates are equally accurate even if the message bits are distributed unevenly over the cover. By adapting the form of Eq. (2.40) to the embedding hypothesis, WS can further be specialised to so-called *sequential embedding*, which means that the message bits are embedded with maximum density (i.e., change rate $1/2 \leftrightarrow p = 1$ in the local segment) in a connected part of the cover. This extension increases the detection accuracy dramatically (by about one order of magnitude), with linear running time still, even if both starting position and length of the message are unknown [128, 133]. Another extension to WS is a generalisation to mod-$k$ replacement proposed in [247].

## 2.11 Summary and Further Steps

If there is one single conclusion to draw from this chapter, then it should be a remark on the huge design space for steganographic algorithms and steganalytic responses along possible combinations of cover types, domains, embedding operations, protocols, and coding. There is room for improvement in almost every direction. So, it is only economical to concentrate on understanding the building blocks separately before studying their interactions when they are combined. This has been done for embedding operations, and there is also research targeted to specific domains (MP [35, 36], YASS [218]) and coding (cf. Sect. 2.8.2). This book places an emphasis on covers because they are relevant and not extensively studied so far.

To study heterogeneous covers systematically, we take a two-step approach and start with theoretical considerations before we advance to practical matters. One problem of many existing theoretical and formal approaches is that their theorems are limited to artificial channels. In practice, however, high-capacity steganography in empirical covers is relevant. So, our next step in Chapter 3 is to reformulate existing theory so that it is applicable to empirical covers and takes account of the uncertainty.

The second step is an experimental validation of our theory: Chapters 4 to 7 document advances in statistical steganalysis. Our broader objective is to develop reusable methodologies, and provide proof of concepts, but we have no ambition to exhaustively accumulate facts. Similarly to the design space for steganographic algorithms, the space of possible aspects of heterogeneity in covers is vast. So closing *all* gaps is unrealistic—and impossible for empirical covers, as we will argue below.

## Remark: Topics Excluded or Underrepresented in this Chapter

Although this chapter might appear as a fairly comprehensive and structured summary of the state of the art in steganography and steganalysis to 2009, we had to bias the selection of topics towards those which are relevant to the understanding of the remaining parts of this book. So we briefly name the intentionally omitted or underrepresented topics as a starting point for interested readers to consult further sources.[51]

We have disregarded the attempts to build provably secure steganography because they fit better into and depend on terminology of Chapter 3. Embedding operations derived from watermarking methods (e.g., the Scalar Costa scheme or quantisation index modulation) have been omitted. Robust steganography has not received the attention it deserves, but little is published for practical steganography. Research on the borderline between covert channels and digital steganography (e.g., hidden channels in games or network traffic [174]) is not in the scope of this survey. Finally, a number of not seriously tested proposals for adaptive or multi-sample embedding functions has probably missed our attention. Quite a few of such proposals were presented at various conferences with very broad scope: most of these embedding functions would barely be accepted at venues where the reviewers consider steganography a security technique, not a perceptual hiding exercise.

---

[51] We also want to point the reader to a comprehensive reference on modern steganography and steganalysis. The textbook by Jessica Fridrich [70] was published when the manuscript for this book was in its copy-editing phase.

# Chapter 3
# Towards a Theory of Cover Models

In this chapter, we develop a theoretical framework to describe the role of knowledge about a cover in the construction of secure steganographic systems and respective detectors as counter-technologies.[1] The objective of such a framework is to give a better understanding of how specific advances in statistical steganalysis presented in Part II can be seen as instances of different types of refinements to a more general cover model.

## 3.1 Steganalyst's Problem Formalised

Before we advance to the actual formalisation, let us briefly elaborate on a dormant simplification that is inherent in all information-theoretic approaches to steganography and steganalysis.

### 3.1.1 The Plausibility Heuristic

In Simmon's model [217], messages exchanged between the communication partners must be inconspicuous, that is, indistinguishable from 'plausible' covers. However, *plausibility* is primarily an empirical criterion and difficult to capture with formal methods. A common approach is to define plausibility in a probabilistic sense, i.e., likely messages are plausible *by definition*. We call this tweak *plausibility heuristic* and note that it goes with an extreme simplification of the original problem statement, although we are not aware of any literature that does not implicitly rely on this heuristic, often without

---

[1] Robustness as a protection goal in the active warden adversary model is not dealt with in this chapter.

saying so.[2] For example, the plausibility heuristic is introduced with the concept of a cover generating oracle in typical game settings for theoretical security analysis. Those games are inspired by established security games in cryptography, but plausibility is not a relevant criterion there. The first explicit game formulation of steganographic security by Katzenbeisser and Petitcolas [117] is reproduced in Appendix E for reference.

It is not the objective of this book to work out all consequences of the plausibility heuristic, so we confine ourselves to a brief sketch of the impact of this simplification. The plausibility heuristic implies that a message is considered as more or less likely with regard to a *universal* probability function. In practice, neither the communication partners nor the warden possess full knowledge of this function (which would correspond to having access to a global oracle), and it is a philosophical question as to whether such a function can exist at all. Instead, it is reasonable to assume that all parties have private concepts of which messages *they think* are plausible or not. Certainly there exists some overlap between the private concepts; otherwise meaningful communication would be impossible. But there are grey areas as well that leave room for misunderstandings. So, in steganographic communication, the communication partners form rational expectations about what the warden might consider as plausible or not, and therefore may deviate from both their own notion of plausibility and the imaginary universal probability function. Conversely, realistic wardens must form expectations about what is plausible for the inmates, considering all available prior knowledge.[3] Of course, each of these expectations can be formulated as conditional probability functions, but those are just models to replace the lack of insight into the respective other parties' cognition.

So, in the previous paragraph, we have identified at least three simplifications that go along with the common plausibility heuristic, namely

1. universal instead of fragmented and context-specific notion of plausibility,
2. simplification of reasoning and cognition by probability functions,
3. ignorance of strategic interaction by anticipation of other parties' likely notion of plausibility.

---

[2] Some expressions of doubt can be found in the revised version of a seminal paper on the information-theoretic approach to steganography. Cachin [34] acknowledges that "assuming the existence of a covertext distribution seems to render our model somewhat unrealistic for the practical purposes of steganography" (p. 54).

[3] One may argue that an omniscient warden is assumed to follow Kerckhoffs' principle [135] and conceive a worst-case warden. For example, Cachin [33, p. 307] states: "Similar to cryptography, it is assumed that Eve [the warden] has complete information about the system except for the secret key shared by Alice and Bob [the communication partners] that guarantees the security," and further clarifies that his notion of 'system' includes all (global) probability distributions, unlike our definition in Sect. 2.1.1 (p. 12). But adherence to Kerckhoffs' principle cannot justify all aspects of the plausibility heuristic.

This list is probably incomplete. Nevertheless, we join all steganography theorists and accept these simplifications in the following to build a tractable formal framework.

### 3.1.2 Application to Digital Steganography

Let $S$ (mnemonic 'scene') be an infinite set of possible natural phenomena, of which digital representations are conceivable. Digital representations are created either from digitisation of real natural phenomena, or they are computer-generated representations to describe arbitrary imaginary natural phenomena.[4] Alphabet $\mathcal{X}$ is a finite set of, possibly ordered, discrete symbols which form the support for digital representations of elements of $S$ in $n$-ary vectors $\boldsymbol{x} \in \mathcal{X}^n$. Without loss of generality, we assume $n$ to be finite and, to simplify our notation, constant.[5]

Applying the plausibility heuristic, we imagine a universal stochastic *cover generating process* Generate : $S \to \mathcal{X}^n$ to define a probability space $(\Omega, \mathcal{P}_0)$ with $\Omega = \mathcal{X}^n$ and $\mathcal{P}_0 : \mathfrak{P}(\mathcal{X}^n) \to [0, 1]$ ($\mathfrak{P}$ is the power set operator). $\mathcal{P}_0$ fulfils the probability axioms,; hence,

$$(\text{i}) : \mathcal{P}_0(\{\boldsymbol{x}\}) \geq 0, \ \forall \boldsymbol{x} \in \Omega, \tag{3.1}$$

$$(\text{ii}) : \mathcal{P}_0(\Omega) = 1, \text{ and} \tag{3.2}$$

$$(\text{iii}) : \mathcal{P}_0(\cup_i \{\boldsymbol{x}_i\}) = \sum_i \mathcal{P}_0(\{\boldsymbol{x}_i\}). \tag{3.3}$$

To simplify the notation, let $\mathcal{P}_0(\boldsymbol{x})$ be equivalent to $\mathcal{P}_0(\{\boldsymbol{x}\})$.

All covers $\boldsymbol{x}^{(0)}$ and stego objects $\boldsymbol{x}^{(1)}$ are elements of $\mathcal{X}^n$, so that the assignment between covers and stego objects is given by function Embed : $\mathcal{M} \times \mathcal{X}^n \times \mathcal{K} \to \mathcal{X}^n$, depending on message $\boldsymbol{m} \in \mathcal{M}$ and key $\boldsymbol{k} \in \mathcal{K}$ (cf. Sect. 2.1.1). As stego and cover objects share the same domain, a passive steganalyst's problem is to find a binary partition of $\mathcal{X}^n$ to classify cover and stego objects Detect : $\mathcal{X}^n \to \{\text{cover}, \text{stego}\}$ based on the probability that a suspect object $\boldsymbol{x}^{(i)}$ appears as a realisation of clean covers, $\text{Prob}(i = 0|\boldsymbol{x}^{(i)})$, or stego objects, $\text{Prob}(i \neq 0|\boldsymbol{x}^{(i)}) = 1 - \text{Prob}(i = 0|\boldsymbol{x}^{(i)})$. These probabilities can be obtained using the Bayes theorem:

---

[4] Note that this definition avoids making a distinction on the philosophical question about whether a (one) cognisable reality exists, as any element of $S$ can either be based on observation of reality or result from human creativity (produced with the assistance of computers).

[5] Finite $n$ can be justified against the backdrop of finite automata used for practical digital steganography: $n$ is bounded by the memory in which the digital representation is stored, and also by the number of operations to process and evaluate it. The requirement of constant $n$ can be relaxed by defining $n \geq n'$ as the maximum cover size and augmenting $\mathcal{X}' = \mathcal{X} \cup \{\bot\}$. Then, trailing elements of $\boldsymbol{x} = \boldsymbol{x}'||\{\bot\}^{(n-n')}, \boldsymbol{x}' \in \mathcal{X}^{n'}$, can be padded with the special symbol $\bot$.

$$\mathsf{Prob}(i = 0 | \boldsymbol{x}^{(i)}) = \frac{\mathsf{Prob}(\boldsymbol{x}^{(i)} | i = 0) \cdot \mathsf{Prob}(i = 0)}{\mathsf{Prob}(\boldsymbol{x}^{(i)})} \tag{3.4}$$

$$= \frac{\mathcal{P}_0(\boldsymbol{x}^{(i)}) \cdot \mathsf{Prob}(i = 0)}{\mathsf{Prob}(i = 0) \cdot \mathcal{P}_0(\boldsymbol{x}^{(i)}) + (1 - \mathsf{Prob}(i = 0)) \cdot \mathcal{P}_1(\boldsymbol{x}^{(i)})}, \tag{3.5}$$

and conversely:

$$\mathsf{Prob}(i \neq 0 | \boldsymbol{x}^{(i)}) = \frac{\mathcal{P}_1(\boldsymbol{x}^{(i)}) \cdot (1 - \mathsf{Prob}(i = 0))}{\mathsf{Prob}(i = 0) \cdot \mathcal{P}_0(\boldsymbol{x}^{(i)}) + (1 - \mathsf{Prob}(i = 0)) \cdot \mathcal{P}_1(\boldsymbol{x}^{(i)})}. \tag{3.6}$$

Evaluating these expressions requires a priori the marginal probability of clean covers $\mathsf{Prob}(i = 0)$ and knowledge of the distribution of all stego objects $\mathcal{P}_1 : \mathfrak{P}(\mathcal{X}^n) \to [0,1]$, which is defined by sets $\Omega, \mathcal{M}, \mathcal{K}$ and functions $\mathcal{P}_0$ and Embed:[6]

$$\mathcal{P}_1(\boldsymbol{x}^{(1)}) = \frac{1}{|\mathcal{M}| \cdot |\mathcal{K}|} \sum_{\boldsymbol{m} \in \mathcal{M}} \sum_{\boldsymbol{k} \in \mathcal{K}} \sum_{\boldsymbol{x} \in \Omega} \mathcal{P}_0(\boldsymbol{x}) \cdot \delta_{\boldsymbol{x}^{(1)}, \mathsf{Embed}(\boldsymbol{m}, \boldsymbol{x}, \boldsymbol{k})}. \tag{3.7}$$

($\delta$ is the Kronecker delta, see Eq. (2.4); the probability axioms, Eqs. (3.1)–(3.3), hold for $\mathcal{P}_1$ as well). A special case exists when the steganalyst knows the message; then $\mathcal{P}_1$ can be replaced by $\mathcal{P}_{1\boldsymbol{m}}$, conditionally on $\boldsymbol{m} \in \mathcal{M}$:

$$\mathcal{P}_{1\boldsymbol{m}}(\boldsymbol{x}^{(1)}) = \frac{1}{|\mathcal{K}|} \sum_{\boldsymbol{k} \in \mathcal{K}} \sum_{\boldsymbol{x} \in \Omega} \mathcal{P}_0(\boldsymbol{x}) \cdot \delta_{\boldsymbol{x}^{(1)}, \mathsf{Embed}(\boldsymbol{m}, \boldsymbol{x}, \boldsymbol{k})}. \tag{3.8}$$

We disregard this special case in the following.

Now it is straightforward to specify function Detect via a threshold parameter $\tau_0$ as

$$\mathsf{Detect}(\boldsymbol{x}^{(i)}) = \begin{cases} \{\text{cover}\} & \text{for} \quad \mathsf{Prob}(i = 0 | \boldsymbol{x}^{(i)}) \geq \tau_0 \\ \{\text{stego}\} & \text{otherwise.} \end{cases} \tag{3.9}$$

This establishes a direct link between $\tau_0 \in [0, 1]$ and the decision-theoretic error measures, false positive probability $\alpha$ and missing probability $\beta$,

---

[6] This relation assumes deterministic embedding functions and uniform distribution of messages and keys. For indeterministic embedding functions, another convolution over the realisations of the random variable in Embed must be considered. Similarly, probability weights for elements in $\mathcal{M}$ and $\mathcal{K}$ can be introduced if the distributions are not uniform.

$$\alpha = \sum_{x \in \Omega} \mathcal{P}_0(x) \cdot \delta_{\mathsf{Detect}(x),\{\mathsf{stego}\}} \tag{3.10}$$

$$\beta = \sum_{x \in \Omega} \mathcal{P}_1(x) \cdot \delta_{\mathsf{Detect}(x),\{\mathsf{cover}\}}. \tag{3.11}$$

The Neyman–Pearson lemma suggests the likelihood ratio test (LRT) as the *most powerful* discriminator between plain covers and stego objects for a given threshold $\tau$ [127, 176]:

$$\Lambda(x^{(i)}) = \frac{\mathcal{P}_0(x^{(i)})}{\mathcal{P}_1(x^{(i)})} \tag{3.12}$$

$$\mathsf{Detect}(x^{(i)}) = \begin{cases} \{\mathsf{cover}\} & \text{for } \Lambda(x^{(i)}) > \tau \\ \{\mathsf{stego}\} & \text{otherwise.} \end{cases} \tag{3.13}$$

'Most powerful' means, for any given false positive probability $\alpha$, a threshold $\tau$ exists so that the resulting missing probability $\beta$ is minimal over all possible detection functions. This theorem holds for tests between two point hypotheses.

While it is easy to write down these formal relations, for practical systems the equations are of limited use for computational and epistemological reasons. We explain this in more detail in the following section.

### 3.1.3 Incognisability of the Cover Distribution

When linking theoretical and practical steganography, there remains a basic epistemological problem: although $\Omega$ is finite, $\mathcal{P}_0$ is incognisable in practice. The probability of each element in $\Omega$ appearing as cover depends on a probability distribution over $\mathcal{S}$, which must be assumed to have infinite support, and as knowledge about it is limited to experience with finite observations, it can never be complete. Consequently, because $\mathcal{P}_1$ depends on $\mathcal{P}_0$ via Eq. (3.7), for non-pathologic functions Embed,[7] $\mathcal{P}_1$ is not cognisable either.

Note that this restricts both computationally unbounded (i.e., theoretical) and bounded (practical) adversaries. The latter may face additional difficulties because evaluating Eq. (3.7) is inefficient in general. Except for artificial channels, for which the cover distribution is defined (see Sect. 2.6.1), both types of adversaries are on a level playing field with regard to the pure steganalysis decision problem. Of course, computationally unbounded adversaries have an advantage if the secrecy of the message $m$, protected by key $k$, is only conditionally secure and the message $m$ contains structure (or is known), so that an exhaustive search over key space $\mathcal{K}$, with $|\mathcal{K}| < |\mathcal{M}|$,

---

[7] $\mathcal{P}_1$ could be known if the output of Embed is independent of the cover input, for example, if Embed overwrites the whole cover signal with the message, or generates an entirely artificial cover.

allows inference on the existence of a secret message. This, however, is a cryptographic problem (resolvable by using unconditionally secure encryption) and not a steganographic one.[8]

## 3.2 Cover Models

The impossibility of finding ground truth on the cover distribution does not prevent practitioners from developing embedding functions and detectors. They thereby rely—explicitly or implicitly—on *models* of the 'true' cover generating process. If the models are good enough, that is, their mismatch with reality is not substantial, this approach is viable.

Before we present our definition of cover models, let us remark that the term 'model' is overused. It can stand for everything from a philosophical standpoint to data structures or single linear equations. It appears as if the scientific language does not allow us to differentiate precisely enough between all aspects subsumable to 'model'. Even within this book, it was not easy to avoid using the term for too many purposes. So let us clarify our notion of a 'model' for covers, which is close to the concept of models in statistics: we think of models as formalisable probabilistic rules, which are assumed to govern a not fully known or understood data generation process. In steganography, this process is the above-introduced cover generating process.

### 3.2.1 Defining Cover Models

Steganographic methods based on imperfect cover models can be secure *unless* the steganalyst uses a more accurate cover model. So the cat-and-mouse race between steganographers and steganalysts can be framed as a race for the best cover model. Early literature on digital steganography reflected on the need for cover models (without naming them so) and options to express them formally:

> "What does the steganalyst know about the cover [...] a priori? He might know it entirely, or a probability space that it is chosen from, or a family of probability spaces indexed by external events (e.g., that a letter 'it is raining' is less likely in good weather), or some predicate about it. This knowledge will often be quite fuzzy." [193, p. 349]

---

[8] Cf. Hopper et al. [105], who show that computationally secure steganography depends only on the existence of secure one-way functions. They define security by statistical indistinguishability and work around the epistemological problem by assuming the existence of a stateful cover generating oracle.

We believe that imposing any functional form for cover models unduly restricts their design space, so we propose a rather broad definition which fits in the framework of Sect. 3.1:

$$\text{Cover models are hypotheses on } \mathcal{P}_0.$$

If we adhere to the notion of hypotheses in the sense of Popper's [197] critical rationalism, a school of philosophical epistemology inspired by discovery in sciences, then cover models can only be falsified, but it is impossible to validate them completely. Cover models that are falsifiable, and thus empirical, but not (yet) falsified are retained until they will actually be falsified. This sequence of trial and error resembles pretty well the chronology of research in practical digital steganography and steganalysis.[9]

A frequently used analogy in epistemology is Newton's theory of universal gravitation, which allows predictions that are sufficiently accurate for many practical purposes. This is so despite the fact that Einstein showed that Newton's theory is merely an approximation in his more general theory of relativity (which itself leaves phenomena unexplained and therefore is just a better approximation of the incognisable ultimate truth).

All practical steganographic algorithms and detectors use cover models—sometimes implicit, sometimes explicit as in model-based steganography [206]—and there are even cases where one cover model is proposed but the suggested embedding function implements a different one. Nevertheless, it seems that developers of embedding functions barely start with the specification of a cover model and perform critical tests thereof against empirical covers. Instead, many new embedding functions have been designed ad hoc, or with the aim of just evading (classes of) existing detectors. We speculate that this is mainly because testing cover models seriously is tedious (it involves large data sets from various sources as representative of typical covers), and sometimes discouraging (because finding counterexamples to falsify a cover model is so easy). As a result, most of the existing evidence on steganographic security draws on what we call 'convenience samples' of covers. This always carries the risk that a proposed method is adapted too much to the properties of the sample used by its developers, and therefore does not generalise. This affects both, proposed embedding functions that have been considered too optimistically as safe and proposed detectors, which often turn out to work much less reliably than claimed when applied to heterogeneous covers.

---

[9] Cryptography was a similarly inexact science before Shannon [213] published the information-theoretic underpinnings in 1949. It is questionable whether such a breakthrough can be expected for steganography as well, since secrecy in terms of relative entropy can be formalised and applied to finite domains, whereas plausibility for general covers cannot. However, for artificial (unlike empirical) covers [33, 34], or covers that can be efficiently sampled [105], steganographic algorithms with 'post-Shannon' properties are possible (but such covers are not always plausible per se). It would be desirable to have a terminological differentiation better than 'theoretical' and 'practical' to distinguish between the two aspects of the steganography problem, which require very different approaches and methodology.

One example where seemingly challenging (because never-compressed) but in fact seriously singular covers led to an (implicit) model for a universal detector with suspiciously good detection performance against LSB matching in the spatial domain is documented in Appendix B. There are many more examples, which we do not mention for brevity.

### 3.2.2 Options for Formulating Cover Models

In Sect. 3.1.2 we have defined $\mathcal{P}_0$ as a function which maps arbitrary subsets of $\Omega = \mathcal{X}^n$ to a probability measure between 0 and 1 while meeting the constraints of the probability axioms. Our definition of cover models in Sect. 3.2.1 defines them as 'hypotheses' on $\mathcal{P}_0$. As there are various ways to state hypotheses on a function such as $\mathcal{P}_0$, we can distinguish several ways to formulate cover models.

- **Direct cover models** Obviously, hypotheses can be formulated by direct assignment of probabilities to individual elements $\boldsymbol{x}_1^{(0)}, \boldsymbol{x}_2^{(0)}, \cdots \in \Omega$, for example, $\boldsymbol{x}_1^{(0)} \mapsto 0.1, \boldsymbol{x}_2^{(0)} \mapsto 0.05$, and so on. Hypotheses for cover models need not assign a probability to *all* elements of $\Omega$. Also, incomplete mappings are valid cover models (as long as they remain falsifiable in theory, i.e., a probability must be assigned to at least one nontrivial subset of $\Omega$). Direct formulations of cover models are impractical for real covers due to complexity and observability constraints. That means it is tedious to assign a value to every possible cover (complexity) and impossible to empirically determine the 'right' value for each cover (observability).
- **Indirect cover models** One difficulty in specifying direct cover models is the large size of $\Omega$ which results from the $n$ dimensions in $\mathcal{X}^n = \Omega$. Indirect cover models reduce the dimensionality by defining projections $\mathsf{Proj} : \mathcal{X}^n \to \mathcal{Z}^k$ with $k \ll n$. Note that the support $\mathcal{Z}$ of the co-domain can differ from alphabet $\mathcal{X}$. Now, probabilities $\pi_1, \pi_2, \ldots$ can be assigned to individual elements $\boldsymbol{z}_1, \boldsymbol{z}_2, \cdots \in \mathcal{Z}^k$, which indirectly specifies the values of $\mathcal{P}_0$ for disjoint subsets of $\Omega$ as follows:

$$\boldsymbol{z} \mapsto \pi \quad \Leftrightarrow \quad \{\boldsymbol{x}^{(0)} | \boldsymbol{x}^{(0)} \in \Omega \wedge \mathsf{Proj}(\boldsymbol{x}^{(0)}) = \boldsymbol{z}\} \mapsto \pi. \qquad (3.14)$$

For example, an indirect cover model along the (simple) intuition that natural covers contain some variation in sample values can be specified by defining $\mathsf{Proj}$ as the empirical variance of the cover signal $\boldsymbol{x}^{(0)}$,

$$z = \mathsf{Proj}(\boldsymbol{x}^{(0)}) = \left(\frac{1}{n} \sum_{i=1}^n \left(x_i^{(0)}\right)^2\right) - \left(\frac{1}{n} \sum_{i=1}^n x_i^{(0)}\right)^2, \qquad (3.15)$$

and setting

$$\mathcal{P}_0\left(\left\{\boldsymbol{x}^{(0)}\mid \boldsymbol{x}^{(0)}\in\Omega\wedge\mathsf{Proj}(\boldsymbol{x}^{(0)})=z\right\}\right)=\begin{cases}0\text{ for }z=0\\1\text{ otherwise.}\end{cases}\qquad(3.16)$$

- **Conditional cover models** Conditional cover models are generalisations of indirect cover models and deal with the combination of the remaining $k$ dimensions of $\boldsymbol{z}\in\mathcal{Z}^k$ after the projection. In practice, there are cases where the marginal distribution of a subspace of $\mathcal{Z}^k$ is unknown (or incognisable), but hypotheses can be formulated on the conditional distribution with respect to *side information* available to the steganographer or steganalyst.

  For example, it is very difficult to stochastically describe the spatial composition of images in general, but it might be easier to formulate a hypothesis on, say, landscape photographs (one expects, on average, a higher blue component or more brightness in the upper part of the image and more prominent edges in the bottom part). Although the true share of such landscape photographs among all photographs is unknown (which impedes the assignment of absolute probabilities in hypotheses about $\mathcal{P}_0$), $\mathcal{P}_{0\mathrm{cond}}$ can be specified for images belonging to the class of landscape photographs. A specific drawback of this method is that the predictive power of the cover model, and thus the accuracy of resulting steganography and steganalysis methods, depends on the availability and accuracy of the condition as side information.

- **Stego models** Finally, stego models can be seen as special cases of indirect cover models, in which Embed serves as a projection rule. Since they depend on a specific embedding function, stego models are most useful in (targeted) steganalysis. They are in fact hypotheses on $\mathcal{P}_1$ which are sometimes easier to formulate intuitively from an analysis of the embedding function. For example, the proof in [233] that embedding algorithm F5 is not vulnerable to the specific chi-squared detector [238] is based on what we would call a stego model. The relation to $\mathcal{P}_0$ is then given by inversion of Eq. (3.7). Even *if* invertible in theory, analytical solutions are intractable in most practical cases. Stego models, like cover models, can be specified directly, indirectly, or as conditional stego models. This may be necessary because Embed, unlike Proj, typically does not reduce the dimensionality of $\Omega$.

Below, in Sect. 3.3, we introduce a class of conditional cover models and methods to estimate the side information which motivate a number of specific improvements of statistical steganalysis methods presented in Part II. Before that, we devote one section to further illustrating examples to explain properties of cover models—in particular the relation to detection performance—for very simple imaginary stego systems and covers.

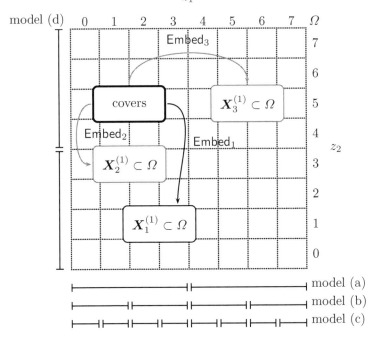

Fig. 3.1: Examples for cover models: 'world view' of projection plane $(z_1, z_2)$ with embedding functions $\mathsf{Embed}_1, \ldots, \mathsf{Embed}_3$ and cover models (a)–(d)

### 3.2.3 Cover Models and Detection Performance (by Example)

In this section we introduce and discuss, step by step, a simple enough example of a small 'world model'. Based on this, we explain differences in knowledge about the world by formulating several further simplified cover models. The aim of this section is twofold. First, we illustrate some of the aspects introduced on a more theoretical level in the previous sections. Second, we elaborate on how cover models can be compared with respect to their detection performance.

Imagine a high-dimensional universe of signals $\Omega = \mathcal{X}^n, n \gg 2$ where there exists a projection $\Omega \rightarrow \{0, \ldots, 7\}^2$ so that all covers $x^{(0)} \in \Omega$ are projected to vectors $Z^{(0)} = (z_1, z_2) \in \{(1, 5), (2, 5)\}$ with equal probability[10] (see Figure 3.1 for a graphical representation).

---

[10] Random cover selection or indeterminacy in the cover generation process motivate the introduction of random variables for the projections of covers.

For now, consider embedding function $\mathsf{Embed}_1 : \Omega \to \Omega$, which embeds a random secret message using a random key. All stego objects $\boldsymbol{X}_1^{(1)}$ resulting from covers $\boldsymbol{X}^{(0)}$ are projected to $\boldsymbol{Z}^{(1)} \in \{(2,1),(3,1)\}$, again with equal probability and, for a specific cover $\boldsymbol{x}^{(0)}$, independently of $\boldsymbol{Z}^{(0)}$. What appears obviously insecure given the 'world view' might have been reasonable from a less informed perspective. The choice of this embedding function could have been motivated by a simple cover model that incorporates less knowledge about the imaginary world than explained so far. For example, the only projection known to the designer of $\mathsf{Embed}_1$ could have been $z_1$, at a lower resolution than possible:

$$\mathsf{Proj}_{(a)}(\boldsymbol{x}) = \begin{cases} 0 \text{ for } z_1 < 4 \\ 1 \text{ otherwise.} \end{cases} \tag{3.17}$$

So the embedding function's cover model corresponds to model (a) in Fig. 3.1 and is specified indirectly as

$$\mathcal{P}_0\left(\left\{\boldsymbol{x} \mid \boldsymbol{x} \in \Omega \wedge \mathsf{Proj}_{(a)}(\boldsymbol{x}) = i\right\}\right) = \begin{cases} 1 \text{ for } i = 0 \\ 0 \text{ otherwise.} \end{cases} \tag{3.18}$$

With model (a) alone, covers and stego objects from $\mathsf{Embed}_1$ are indistinguishable. If both steganographer and steganalyst were limited to this model, then secure steganography would be possible, as stated in early information-theoretic literature on steganographic security:

> "A secure stegosystem requires that the users and the adversary share the same probabilistic model of the covertext" [34, p. 54]

In light of our theory, this proposition is deemed too restrictive, because the requirement to share the *same* model is a sufficient condition for secure steganography, but not a necessary one. The steganalyst may well use a *different* model as long as the steganalyst does not possess a *better* cover model than that implemented in the embedding function. The term 'better' suggests that there must be an order of cover models for a fixed cover distribution $\mathcal{P}_0$ and embedding function. And we argue that the most reasonable order is determined pragmatically by the models' ability to discriminate between covers and stego objects, i.e., their detection performance.[11]

It is evident that model (b) in Fig. 3.1 is superior to model (a):

---

[11] Aside from detection performance, model complexity is probably the second most relevant criterion which could serve for ordering cover models. Adhering to the principles of *Occam's razor* or the *lex parsimoniae*, one could reason about discounting detection power by a penalty for overly complex models. Although we indeed find justification for taking model complexity into account (for normative reasons as well as statistical and epistemological ones [197]), we omit this aspect here because finding appropriate complexity metrics is difficult and aggregation methods to combine them with performance metrics are largely unexplored.

$$\mathsf{Proj}_{(b)}(\boldsymbol{x}) = \begin{cases} 0 \text{ for} & z_1 < 2 \\ 1 \text{ for } 2 \le z_1 < 4 \\ 2 \text{ for } 4 \le z_1 < 6 \\ 3 \text{ otherwise.} \end{cases} \tag{3.19}$$

In an interactive game for security analysis (cf. Appendix E) with $\mathsf{Prob}(i = 0|\boldsymbol{x}^{(i)}) = \mathsf{Prob}(i = 1|\boldsymbol{x}^{(i)}) = \frac{1}{2}$, the conditional probabilities for a suspect object being a *cover* and the respective detector outputs are given as follows:

$$\mathsf{Prob}\left(i = 0|\mathsf{Proj}_{(b)}(\boldsymbol{x}^{(i)}) = k\right) = \begin{cases} 1 \text{ for } k = 0 \text{ (detector output `cover')} \\ \frac{1}{3} \text{ for } k \ge 1 \text{ (stochastic output, } 66.\overline{6}\% \text{ `stego').} \end{cases} \tag{3.20}$$

Unlike model (a), where covers and stego objects were indistinguishable, model (b) with the suggested detector rule detects $1 - \beta = 66.\overline{6}\%$ of the stego objects created by $\mathsf{Embed}_1$ correctly while wrongly classifying $\alpha = 33.\overline{3}\%$ of covers as stego objects (AUC=$1/2$).[12]

Further refinements of the model, e.g., $\mathsf{Proj}_{(c)}(\boldsymbol{x}) = z_1$, yield even better detection performance. A steganalyst equipped with model (c) can achieve a detection rate of $1 - \beta = 75\%$ at a false positive rate of $\alpha = 25\%$ against embedding function $\mathsf{Embed}_1$ (AUC=$3/4$).[13] This is so because the ranges of cover and stego objects of $\mathsf{Embed}_1$ overlap on this dimension. The differences in detection performance can also be visualised as ROC curves in Fig. 3.2. The corresponding analytical derivation of the curves and related performance metrics is documented in Appendix F. Observe that perfect separation (i.e., $\alpha = \beta = 0$) is not possible with information about $z_1$ alone.

Even worse, if the steganographer learns about the existence of model (c), he could try to adapt his embedding function to something like $\mathsf{Embed}_2$ (see Fig. 3.1), stego objects of which are indistinguishable from covers (i.e., the function is secure) with respect to model (c). As models (a) and (b) are *nested* in the more general model (c), $\mathsf{Embed}_2$ is also secure against steganalysts using models (a) and (b). A model is called nested in another model if all possible detectors of the former can also be built from the latter. Hence, hierarchies of nested models can be put in a total order from simple (worse) to more general (better) models. A corollary of sorting cover models by detection performance is that the order is dependent on a specific embedding function, or, more precisely, the steganographer's (implicit) cover model.

Unfortunately, a total order in nested models is not enough to theoretically underpin the proposition that steganography is secure unless the steganalyst finds a 'better' model. In fact, the 'better' model of the steganalyst may be ridiculously simple if it evaluates just *another dimension* not considered in

---

[12] We report the trade-off for the EER. Other combinations are possible as well, e.g., $1 - \beta = 100\%, \alpha = 50\%$, but the false positives cannot be reduced to zero while maintaining positive detection rates $1 - \beta$.

[13] Again, other detector outputs yield different rates within the limits $1 - \beta = 100\%, \alpha = 50\%$ to $1 - \beta = 50\%, \alpha = 0\%$.

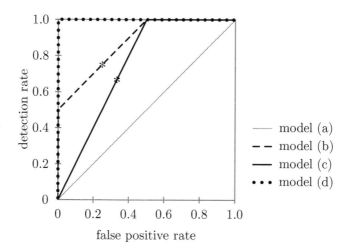

Fig. 3.2: ROC curves: performance of detectors based on cover models (a)–(d) against $\mathsf{Embed}_1$. ERRs for nontrivial curves are marked with symbol '$*$'

the most sophisticated model employed by the steganographer. An example is given in model (d), which evaluates dimension $z_2$ only coarsely but effectively; both $\mathsf{Embed}_1$ and $\mathsf{Embed}_2$ can be detected perfectly:

$$\mathsf{Proj}_{(d)}(\boldsymbol{x}) = \begin{cases} 0 \text{ for } z_2 < 4 \\ 1 \text{ otherwise.} \end{cases} \tag{3.21}$$

Conversely, if the steganographer had used an embedding function that is secure under model (d), this does not imply that it is secure against steganalysts equipped with detectors based on dimension $z_1$ (models (a)–(c)). $\mathsf{Embed}_3$ in Fig. 3.1 illustrates a counterexample.

As $n \gg k$, other—read linearly independent—dimensions for analysis are abundant, and we are not aware of attempts to prove that a deterministic embedding function (unlike the inefficient[14] rejection samplers [3]) does not alter the statistics of arbitrary projections of $\mathcal{X}^n$. This is why we can tentatively conjecture that provably secure and efficient high-capacitysteganography in empirical covers is probably unachievable.[15]

---

[14] Obstacles in constructing efficient rejection samplers have been studied by Hundt et al. [108]. Papers referring to 'efficient' provable schemes in their title, such as [138] for SKS and [150] for PKS, usually describe algorithmic tweaks in the coding layer, but do not achieve reasonable embedding capacities for practical covers.

[15] The attribute 'provably' in the title of [220] is misleading. The proposed embedding function restores first-order statistics similarly to [198] or [62]. It is insecure against detectors that use more elaborate cover models than histograms.

Universal detectors serve as another example of where 'better' models of the steganalyst are generally not just generalisations of the steganographer's model. Trained universal detectors (i.e., the feature set, classification method and parameters) are in fact empirically calibrated indirect cover models, and the feature extraction corresponds to function Proj [127, 188]. Also, many targeted detectors employ unnested cover models because it is usually easier to catch the steganographer on a dimension that has not been considered. As a common principle in security engineering, the adversary is free to choose the weakest link [209, Chapter 8, among many others].

## 3.2.4 Summary and Motivations for Studying Cover Models

Summing up the section on cover models so far, we have proposed a definition of cover models as hypotheses about the empirical cover distribution. And we have presented a system to differentiate ways for formulating such hypotheses. Then, a stylised example of a cover universe that can be reduced to two dimensions has been introduced to illustrate the relation between the choice of cover models by steganographers and steganalysts, and its implication on detection performance. Thereby we have emphasised, in exposition and by example, what we understand by 'better' cover models.

As the success of steganography or steganalysis is determined by the quality of the cover model, a systematic approach for reasoning about cover models is preferable to ad hoc efforts. Explicit cover models (as hypotheses) can be tested against true covers independently from steganographic systems. This is not only more efficient, because no stego channel adds to the overhead and results are reusable for various specific methods, but most likely also more effective, as large problems (e.g., secure steganography) can be broken down into smaller, more tractable ones (e.g., appropriateness of a specific cover assumption). Better insight into cover models helps make steganalysis more reliable in at least two ways:

1. Knowledge about the (implied) cover model of a specific embedding function facilitates an analytical approach to derive detection functions that deliberately exploit the mismatch between the cover model of the embedding function and true empirical covers.
2. Insight into (implied) cover models of existing detectors may help to identify conditions under which the model works better and others under which it is wrong. If these conditions can be evaluated for *individual* detection decisions for a given object under analysis, then the detector parameters, including the decision threshold, can be adapted to the particular object. In the extreme case, different detectors can be applied depending on which

cover model fits better for a specific input. On average, this leads to tighter bounds for the error rates and to better detection performance.

Conversely, also the development of steganographic algorithms can benefit from a better understanding of cover models. Broadly corresponding to the two aspects in steganalysis, we can say the following for steganography:

1. Knowledge of reasonably valid cover models helps us to construct better, possibly cover-specific, embedding techniques. One example for such an approach is perturbed quantisation steganography, which deliberately mimics the—somewhat singular, but in practice not uncommon—artefacts of JPEG double-compression [83].
2. Insight into (implied) cover models of existing embedding functions may help us to identify conditions under which the model fits better and others under which it fits less well. This information could be used for an educated selection of suitable covers or for an automatic rejection of insecure covers [137, 193].[16]

Finally, studying cover models may yield to spillovers beyond the area of steganography and steganalysis. Digital watermarking, multimedia forensics and certain areas of biometric security all share with steganography and steganalysis both the need for cover models and the security perspective. With respect to cover models, a security perspective means that special emphasis is laid on the guaranteeable properties in the $\varepsilon$-worst (or $\varepsilon$-best) case for the defender (or adversary). Other (non-security) domains dealing with cover models as well, such as multimedia source coding and computer vision, focus rather on the average case.

As advances on cover models are useful in many aspects, we continue with a proposal for a novel way of looking at a class of cover models.

## 3.3 Dealing with Heterogeneous Cover Sources

One reason why $\mathcal{P}_0$ is so difficult to model and approximate is the fact that typical cover formats are so general that objects from a large number of different sources can be stored and transmitted. Each of the cover generating processes of various sources is governed by specific physical mechanisms, resulting in different statistical characteristics of the so-obtained digital representations. It is even difficult to assess the amount of variation between cover sources, as the unit of analysis to estimate statistical properties would

---

[16] Note that cover preselection itself biases the cover distribution and degrades the theoretical security of the stego system. However, this weakness can be exploited by the steganalyst only asymptotically with sequential steganalysis of many objects on the channel. In practice, the short-term benefits of not being caught *today* might well outweigh the small theoretical accumulation of extra evidence in the long run.

be the source and not the individual cover. Common cover data sets consist of many covers from a single (homogeneous) source, whereas controlled sampling from a large number of heterogeneous sources is much more costly. Consequently, many efforts to refine cover models effectively come up with more specific models for particular sources, but what is really needed is a unified approach to deal with heterogeneous cover sources in general.

Note that we use an informal definition of *heterogeneity*: variation in a random variable is too large to ignore the error of approximating its realisations by fixed parameters of the distribution function (e.g., moments, if applicable). Multi-modality (i.e., local maxima of the distribution function) or non-negligible dispersion (e.g., heavy tails) are common causes and signs of heterogeneity. Of course, the criterion of what is acceptable depends on the application and, in the case of steganalysis, is connected to the achievable security or detection performance.

Against the backdrop of our theory, we argue that heterogeneity can be best tackled with conditional cover models (see Sect. 3.2.2), which combine several more specific cover models for homogeneous subsets of heterogeneous cover sources, possibly in a hierarchical manner. The idea is motivated by the observation that measuring unconditional dependencies in high-dimensional empirical covers is often impractical, but conditional versions of $\mathcal{P}_0$ are (more) tractable. After introducing a motivating example, we recall the basic principle of mixture distributions as an elegant and statistically well-founded way to deal with heterogeneity. Then we explain how this concept can be applied to cover models. Our goal is still to build a general conceptual framework, which can be tailored to specific applications.

Let us briefly introduce an example to which we will refer in the course of this section. It is a standard problem in steganalysis of greyscale images in spatial domain representation. There is a wide range of possible sources for this type of cover, but precise enough cover models for the general case are unknown. However, in reality, some covers of this class have previously been compressed with a lossy compression algorithm, such as JPEG. For those covers, much more precise cover models are at hand [148] (further refined for specific contents in [147]) and can be exploited to the advantage of steganalysis [75]. So if JPEG pre-compressed images can be identified as such, more appropriate cover models can be employed to improve steganalysis. Identification of cover histories can either happen through side information, as mentioned in Sect. 3.2.2, or by estimating the most likely origin from the image itself [55]. Hence, although we do not know the distribution of a specific pixel conditionally on all other pixels in general, we can compute a (low-dimensional) summary measure to identify previously JPEG-compressed images and use this knowledge to verify if local (again, low-dimensional) dependencies between pixels are consistent with the JPEG history. This way, different cover properties (origin, preprocessing, etc.) quite naturally lead us to the more abstract notion of modelling the probability space $(\Omega, \mathcal{P}_0)$ with mixture models.

### 3.3.1 Mixture Distributions

To recapitulate the ideas behind mixture distributions, consider the simplest possible mixture model. Let $Y$ and $Z$ be independent random variables with arbitrary probability distributions and let $H \sim \{0, 1\}$ be a binary random variable independent of $Y$ and $Z$. Now consider a data generating process for a dependent random variable $X$:

$$X = HY + (1 - H)Z. \tag{3.22}$$

$X$ is said to follow a *mixture distribution* because realisations of $Y$ and $Z$ are mixed in $X$ depending on $H$. Typically, realisations of $X$ are observable data, but the mixture structure hides whether a specific realisation of $x$ is actually a realisation $y$ of $Y$ (if $h = 1$) or $z$ of $Z$ (if $h = 0$). Realisations of $H$ are not observable and are called *hidden data* in the statistics literature. However, we do not use this term in this book and prefer *unobservable random variables* or *unobservable realisation*, respectively, to avoid confusion with the secret message, which is the 'hidden data' in the context of information hiding.[17]

Without going into unnecessary detail, it is sufficient to know that there exist methods to check whether the mixture model is *identifiable*, that is, to estimate the marginal distributions of $Y$, $Z$ and $H$ from a number of data points $x_1, \ldots, x_N$. The well-known expectation maximisation (EM) algorithm [47] estimates the parameters by updating beliefs about the unobservable realisations $h_1, \ldots, h_N$ and parameter estimates for $Y$, $Z$ and $H$, alternatingly. Other estimation procedures include Monte Carlo approaches or—for some classes of distributions—spectral methods. These methods are preferable when the EM algorithm converges slowly or not at all.

### 3.3.2 The Mixture Cover Model

The application of the mixture framework to cover models is quite straightforward. The heterogeneity in cover sources (origin, preprocessing, etc.) can be modelled as different cover distributions $\boldsymbol{X}_{H_1}^{(0)} \sim (\Omega, \mathcal{P}_{H_1})$, $\boldsymbol{X}_{H_2}^{(0)} \sim (\Omega, \mathcal{P}_{H_2})$, $\ldots$, $\boldsymbol{X}_{H_k}^{(0)} \sim (\Omega, \mathcal{P}_{H_k})$. The distribution of all empirical cover is thus a mixture of all $k$ distributions, and realisations of the discrete random variable $H \in \{1, \ldots, k\}$ are unobservable:

$$\boldsymbol{X}^{(0)} = \sum_{i=1}^{k} \delta_{H,i} \, \boldsymbol{X}_{H_i}^{(0)} \sim (\Omega, \mathcal{P}_0). \tag{3.23}$$

---

[17] Our notion of 'unobservable' is borrowed from statistics and differs from the definition in [192]. It particular, it does consider any relation between senders and recipients of messages.

Two modifications of this model come to mind in practice. First, the number of different source distributions $k$ is often unknown, so we reduce the domain of $H$ to the binary case: either the conditional cover model is appropriate for a particular cover ($h = 1$, e.g., the image has previously been JPEG-compressed) or not ($h = 0$, i.e., all other possible preprocessings, including no preprocessing at all). The resulting distribution model is closer to the simplest possible case in Eq. (3.22),

$$\boldsymbol{X}^{(0)} = H\boldsymbol{X}_H^{(0)} + (1 - H)\boldsymbol{X}_{\overline{H}}^{(0)}. \tag{3.24}$$

To continue our example, let $h = 1$ for covers that have been pre-compressed with JPEG, and $h = 0$ if not. The unknown marginal distribution of $H$ is the empirical distribution of JPEG pre-compressed versus never-compressed images. A refined cover model for JPEG pre-compressed images could be employed conditional on $h = 1$.

The second modification follows from the standard model in steganalysis, namely that cover models describe the covers $\boldsymbol{X}^{(0)}$, but security is determined by the detectability of the resulting stego object $\boldsymbol{X}^{(1)}$, i.e., after applying the embedding function. This means that the method to find $\hat{h}$, an estimate of the unobservable realisation $h$ of $H$, should be (asymptotically) invariant to the embedding function Embed. Formally,

$$|\mathsf{Prob}(h = \hat{h}|\boldsymbol{x}^{(0)}) - \mathsf{Prob}(h = \hat{h}|\boldsymbol{x}^{(1)})| \text{ is negligible } \forall\, h. \tag{3.25}$$

The implications of the mixture framework on detection performance is illustrated in Fig. 3.3, which extends the example developed in Sect. 3.2.3. The cover distribution is a mixture of two distinct (in the projected plane) random variables, $\boldsymbol{X}_H^{(0)}$ for JPEG pre-compressed covers and $\boldsymbol{X}_{\overline{H}}^{(0)}$ for never-compressed covers. Assume that the effect of the embedding function in the projected plane is as depicted. A steganalyst equipped with either model (a) or model (d)[18] cannot distinguish between clean covers of either type and stego objects (AUC = 0). However, model (d) can be used to distinguish between pre-compressed and never-compressed covers on dimension $\hat{h} = h \propto z_2$, so that a steganalyst who has access to both models can base a decision rule on model (a) *conditional* on the realisation $h$ of $H$ and thus distinguish between covers and stego objects perfectly (AUC = 1).[19]

At first sight, the intertwined transformation of Embed in the projected plane of this example may appear artificial and too trivial. But this is not a serious shortcoming. Aside from a slight exaggeration in this example, namely the switch between perfect indistinguishability and perfect detectability, similar situations are quite common in practice. Due to the high dimensionality

---

[18] For simplicity, we omit the refined models (b) and (c) in this example.

[19] Owing to the simplicity of this particular example, an inverse condition is possible as well. In practice, the cover model is more fuzzy and the unobservable random variable still discrete (but possibly subject to estimation errors, i.e., $\mathsf{Prob}(h \neq \hat{h}) > 0$).

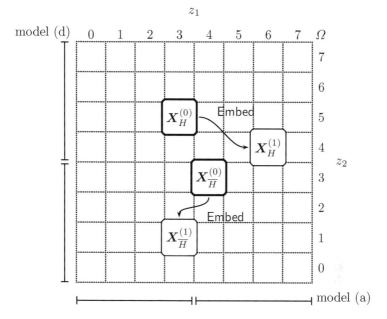

Fig. 3.3: Example for mixture cover models and detection performance

of empirical covers, it is often not very difficult to find appropriate projections to separate different cover sources and select on average the most suitable cover model adaptively.

Note that numerous factors may govern the mixture of covers in practice: $\mathcal{P}_0$ evolves over time with innovation in the development of acquisition devices, standards, software, behaviour, conventions and communication situations. Conditional cover models appear to be the only possibility to (partly) deal with this complexity.

## 3.4 Relation to Prior Information-Theoretic Work

Our theoretical framework is not unprecedented in the literature. Many of the observations formalised in Sect. 3.1 as well as certain aspects of Sects. 3.2 and 3.3 have appeared explicitly or implicitly ('between the lines') in existing work. In this section, we make up for the missing link to prior art and discuss how our framework fits into the most important theoretical works on steganography and steganalysis.

## 3.4.1 Theoretical Limits

Cachin's [33, 34] information-theoretic definition of steganographic security is widely accepted in the literature.[20] He describes the steganalysis problem as a statistical hypothesis test based on the observation of a given suspect object $\boldsymbol{x}^{(i)}$ with

- $H_0 : \boldsymbol{x}^{(i)}$ is generated according to $\mathcal{P}_0$, i.e., the object is a clean cover ($i = 0$), and
- $H_1 : \boldsymbol{x}^{(i)}$ is generated according to $\mathcal{P}_1$, i.e., the object is a stego object ($i = 1$).

The decidability of the hypothesis test can be related to the KLD between the cover and the stego object distributions,

$$D_{KL}(\mathcal{P}_0, \mathcal{P}_1) = \sum_{\boldsymbol{x} \in \Omega} \mathcal{P}_0(\boldsymbol{x}) \log \frac{\mathcal{P}_0(\boldsymbol{x})}{\mathcal{P}_1(\boldsymbol{x})} \le \varepsilon, \tag{3.26}$$

to obtain a definition of $\varepsilon$-secure steganography. Perfectly secure steganography is a special case with $\varepsilon = 0$. Note that, unlike the simple measure in Eq. (2.2), perfect steganography by this definition is secure against *all* possible detectors. The fact that deterministic processing cannot increase the KLD of two distributions allows us to derive bounds for the error rates $\alpha$ and $\beta$, namely

$$D_{bin}(\alpha, \beta) \le \varepsilon \qquad \Leftrightarrow \qquad \beta \ge 2^{-\varepsilon} \text{ for } \alpha = 0. \tag{3.27}$$

Note that Cachin uses the term 'hypothesis' from a statistical background. This should not be confused with our notion of hypotheses in the definition of cover models. Both kinds are in fact conjectures, but Cachin's hypotheses do not necessarily refer to an incognisable reality. They merely postulate the presence of steganographic modifications in observed acts of communication, whereas ours refer to the probability distribution of observing a specific cover *in reality*. Therefore, our notion of hypotheses always concurs with the notion in Popper's theory [197]. It is easy to conceive pathological channels where Cachin's hypotheses are either not falsifiable (perfect steganography in artificial channels: $D_{KL}(\mathcal{P}_0, \mathcal{P}_1) = 0$) or verifiable (perfectly insecure steganography: $\Omega^{(0)}, \Omega^{(1)} \subset \Omega \wedge \Omega^{(0)} \cap \Omega^{(1)} = \varnothing$). Though this is not a serious concern, as in all practical situations with empirical covers, both types of hypotheses fulfil Popper's criteria.

---

[20] The critique of Chandramouli and Memon [38] to the work of Cachin [33] misses the point in that marginal probability distributions of individual samples are inserted in the security definition instead of $n$-dimensional joint distributions of cover objects. The alternative definition by Zöllner et al. [258] confuses secrecy of the message content with security of the existence of *any* message, as pointed out in [34, 253], and has therefore been abandoned in subsequent theoretical work. The concept of dividing a cover into a deterministic and indeterministic part discussed in [258], however, is useful despite the problematic security definition. A less formal presentation of this idea appeared in [63].

Alternative distance measures have been proposed for the definition of $\varepsilon$-secure steganography.[21] Katzenbeisser and Petitcolas [117] as well as Hopper et al. [105] employ the statistical distance in their formal security proofs,

$$\mathsf{D_{SD}} = \frac{1}{2} \sum_{x \in \Omega} |\mathcal{P}_0(x) - \mathcal{P}_1(x)| \leq \varepsilon_{\mathsf{D_{SD}}}. \tag{3.28}$$

Barbier and Alt [9] discuss a weaker security measure based on statistical distance that can be obtained by specialising the measure to discrimination statistics of specific detectors (i.e., specific projections Proj of indirect cover models). Their paper also puts forward (not very surprising) arguments that zero statistical distance between $\mathcal{P}_0$ and $\mathcal{P}_1$ implies security against specific detectors, that is, zero statistical distance between $\mathsf{Proj}(\mathcal{P}_0)$ and $\mathsf{Proj}(\mathcal{P}_1)$, but not vice versa. Zhang and Li [253] define steganographic security by the variational distance between $\mathcal{P}_0$ and $\mathcal{P}_1$,

$$\mathsf{D_{VD}} = \max_{x \in \Omega} |\mathcal{P}_0(x) - \mathcal{P}_1(x)| \leq \varepsilon_{\mathsf{D_{VD}}}. \tag{3.29}$$

However, neither statistical nor variational distance has such a clear link to the error rates as the KLD. The maximum operator in the definition of the variational distance further indicates that this measure is only suitable for the analysis of maintainable security in the worst case. In practice, if $\Omega$ is large, a small number of insecure covers does not harm a lot if the probability that such covers occur is negligibly small.

## 3.4.2 Observability Bounds

Cachin [33] acknowledges that the abstraction to indistinguishability between cover and stego object distributions, $\mathcal{P}_0$ and $\mathcal{P}_1$, respectively, fails to address the "validity [...] [of] a model for real data" (p. 307). Due to the incognisability of $\mathcal{P}_0$ for empirical covers, the information-theoretic bounds of Eqs. (3.26) to (3.29) cannot be calculated in practice. What matters for practical steganography is the empirical distinguishability between (a finite set of) observed stego objects and an estimated (from finite observations) cover distribution. So the practical steganalyst's decision is always based on *incomplete information*. Even if Kerckhoff's principle [135] is interpreted in a very strict sense, in assuming that the steganalyst has supernatural knowledge of $\mathcal{P}_0$, the number $N$ of observed suspect objects may be too small to allow statistical inference on the dissimilarity of sequence $x_1^{(i)}, \ldots, x_N^{(i)}$ with sufficient certainty. So, observability bounds appear to be a much more relevant constraint to practical

---

[21] All definitions have in common that $\varepsilon = 0$ denotes perfectly secure steganography, but less secure stego systems yield different $\varepsilon > 0$ depending on the definition.

steganographic security (and reliability of steganalysis) than the information-theoretic limits. In our initial presentation of components of an adversary model (Sect. 2.5), observability bounds can be subsumed under 'knowledge' of the adversary.

Interestingly, these bounds are not stable. Anderson [4] already mentioned in his seminal paper of 1996 that due to the central limit theorem (CLT), the steganalyst's estimated statistics of the distribution of suspect objects become more accurate as communication continues. So if $\mathcal{P}_0$ and $\mathcal{P}_1$ are not identical, then evidence accumulates and a channel's secure capacity can be gradually 'worn out' over time.

This result has been known in the area of covert channels in operating systems for long [4], but steganography scholars did not study channel wear-out and observability bounds until recently. The only vaguely related source we are aware of is an analysis of an artificial example channel by Chandramouli and Memon [38]. The authors model steganalysis as the difficulty of distinguishing between samples of two Gaussian distributions with different means. Equation (9) of [38], albeit interpreted by the authors as a capacity result, can be solved for the cover size and then be seen as a specific formal instance of the diminishing capacity phenomenon, which Anderson outlined only informally.

The topic has been reopened by Ker [124], who comes to an interesting conclusion for steganographic capacity, though his formal argument is currently limited to the special case of pooled steganalysis (cf. Sect. 2.5.1) and some special instances of artificial channels [58]. Ker [124] postulates and supports with experimental evidence a *batch steganographic capacity theorem*. The theorem says that if the embedded message length increases faster than the square root of the total cover size, then asymptotically the steganalyst will accumulate enough evidence to detect the use of steganography with certainty.[22] Of course, the initial evidence and speed of the convergence may differ, depending on the relative quality of the cover models.

Also related to observability bounds are the most recent works of Ker [127, 131] and Pevný and Fridrich [188], who deal with the *estimation* of information-theoretic distance metrics between empirical distributions for the purpose of steganalysis benchmarking and, more recently, for optimising embedding operations [132]. Both teams use an indirect cover model to reduce the dimensionality of $\Omega$, so effectively $\mathcal{P}_0$ and $\mathcal{P}_1$ are approximated by distance metrics between samples of $\mathsf{Proj}(\boldsymbol{X}^{(0)})$ and $\mathsf{Proj}(\boldsymbol{X}^{(1)})$. The results were too new to be considered in the dissertation on which this book is based. So, little can be said about the usefulness of these metrics for systematic improvements of cover models, but the prospects appear promising.

---

[22] That is, with asymptotically small error rates for $\alpha$ and $\beta$.

## 3.4.3 Computational Bounds

Consider a situation where the steganalyst has enough information available to calculate a particular test statistic, but the calculation requires more computing cycles (or memory) than available. As a result, the stego system is conditionally secure against this particular computationally bounded adversary. Computationally secure steganography has been studied by Hopper et al. [105], who prove that the existence of secure steganography (in covers that can be efficiently sampled conditionally on the history) is implied by the existence of secure one-way functions.[23] Their proof by reduction shows that detecting steganography with asymptotic advantage $\varepsilon_{D_{SD}} > 0$ implies the prediction of bits of a pseudorandom function with higher probability than random guessing. The system is only computationally secure, because every pseudorandom function can be inverted by tabulating all possible states. The generalisation to one-way functions builds on another result in cryptographic theory by Hastad et al. [99]. On the same basis, Ahn and Hopper [3] and, independently, Backes and Cachin [8] extend the analyses to PKS. Since they need asymmetric cryptographic primitives, the link to complexity-theoretic security is even more evident and, depending on which crypto functions are actually used for the primitives, the security also depends on the validity of assumptions on the hardness of certain underlying mathematical problems.

While the hierarchical order of information-theoretical and computational bounds is well understood, we are not aware of literature in the field of steganography that discusses the relation between computational and observability bounds. Note that depending on the computation model, observability bounds can be expressed as computational bounds (each observation costs cycles and memory). For example, it is common practice in constructions of secure steganography to count calls to the rejection sampler oracle as (linear) operations in the complexity of the algorithm. But the availability of a sufficiently long sequence from these oracles is usually not questioned. This is problematic as access to empirical oracles may be limited (or costly) and sampling from artificial channels is not always efficient [108]. Therefore, we argue that it makes sense to distinguish between computational and observability bounds, because in general, neither can an observation constraint be compensated with more computations, nor the converse.[24]

---

[23] Information-theoretically (unconditional) secure steganography is possible in channels with known distribution, as first shown by Cachin [33]. Knowledge of the distribution implies the possibility to sample; however, according to Hundt et al. [108] this is not always efficient.

[24] Computations can replace observations only for fully specified channels; observations help overcoming computational bounds only if so-called *oracles* with precomputed solutions are observable. Both are strong assumptions (the latter is explicitly ruled out in [105]; see the remark on p. 81 of [105]).

## 3.4.4 Applicability of the Theory of Cover Models

The purpose of this section is to recall and clarify under exactly which conditions the theory of cover models developed in this chapter is needed. This is not very easy, as a number of partly independent attempts in the literature to formalise steganographic security makes it difficult to maintain a good overview of the relation of various parameters in the design space for steganographic systems. Table 3.1 tries to remedy this situation and summarises the essence of the previous subsections. The idea is to contrast *assumptions on the cover source* with *assumptions on the limits of the adversary* in a $2 \times 3$ table [19]. The table distinguishes artificial (cf. Sect. 2.6.1) and empirical (cf. Sects. 2.6.2 and 2.6.3) covers in columns. Adversary assumptions are listed in rows and should be interpreted as follows.

Table 3.1: Classification of approaches to secure steganography

| Adversary assumption | Cover assumption | |
|---|---|---|
| (capacity bound) | **artificial** | **empirical** |
| **information-theoretic** | **possible** | **impossible** |
| (infimum of joint entropy of random variables and key) | but finding a secure embedding function can be NP-hard [33, 34] | |
| **complexity-theoretic** | **possible** | **possible** |
| $\left(\frac{\text{cover size } n}{\text{min. sampling unit}}\right)$ | but embedding is not always efficient [105, 108] | if sampling oracle exists (observability) [105] |
| **heuristic** | **likely insecure** | **possible** |
| (0 as cover size $n \to \infty$, asymptotically square root of $n$ [58, 124, 134]) | since steganalyst knows the cover distribution [135] | security depends on the relative accuracy of steganographer's and steganalyst's cover model |

**Information-theoretic** security refers to unconditional security, where the indistinguishability between cover and stego distribution can be proven with information theory. Of course, this is only possible for artificial channels and finding good embedding and extraction functions can be very difficult for general channel models. Cachin [34] shows that this corresponds to solving the NP-hard partition problem. Nevertheless, simple instances of the partition problem can be found for simple example channels (random numbers, public keys, etc. [4, 33, 117]). In combination with information-theoretically secure authentication codes, security even against active adversaries can be achieved

[216]. But the plausibility of such simple instances of artificial channels is highly questionable in practice.

Interestingly, **complexity-theoretic** steganography (to which we subsume also methods that rely on 'well analysed' mathematical assumptions) is possible both in artificial and empirical covers. The trick is to allow constructions that have access to an oracle which can sample from the channel. For theoretical channels, a sampler can always be constructed (but may be computationally complex), whereas finding good samplers for empirical covers is more difficult. The main obstacle is that the *minimum sampling unit* (measured in the number of symbols of $\mathcal{X}$) determines the capacity of the resulting steganographic system. While it is not too difficult to draw independent samples from an unordered series of digital photographs (i.e., the minimum sampling unit is the size of an entire image), it is *very* difficult to draw the intensity of the pixel at position $(u, v)$ conditionally on all pixels $(i, j), 1 \leq i < u, 1 \leq j < v$ (minimum sampling unit 1 if $\mathcal{X}$ is the range of intensity values). Obviously it makes a difference if the capacity is about 1 bit *per image* in the former case or *per pixel* in the latter.

Note that we report the capacity class only as an approximate order of magnitude. Within certain limits, capacity can be traded off against embedding complexity through coding (Sect. 2.8.2) or algorithmic improvements of the oracle query strategy [138, 150].

If no security proofs are known for a given embedding method (heuristic class), and this method is applied to artificial covers, then we have to assume that the adversary knows the specification of $\mathcal{P}_0$ [135]. Given enough stego objects, the adversary can identify deviations of $\mathcal{P}_1$ from the theoretical distribution. Since it is very unlikely that a method which is not designed to be provably secure accidentally preserves $\mathcal{P}_0$, we label this cell as 'likely insecure'.

Nevertheless, heuristic approaches dominate in practical steganography because they are the only ones that allow high embedding capacities for empirical (i.e., plausible) covers. We cite the square root law as guiding principle in the table. Not surprisingly, our notion of cover models and the idea that steganographic security depends on the relative quality of the steganographer's and steganalyst's cover model, apply only to the bottom-right (shaded) field of Table 3.1. All embedding operations discussed in Sect. 2.7, which modify samples at a lower granularity than the sampling unit, belong exclusively to this security class. They do not belong to the class of complexity-theoretic adversary assumptions because the effect of repeated application of embedding operations on individual samples $x_1^{(0)}, \ldots, x_n^{(0)} \mapsto x_1^{(1)}, \ldots, x_n^{(1)}$ on the likelihood $\mathcal{P}_0(\boldsymbol{x}^{(1)})$ of the resulting cover object is analytically intractable if interdependencies between samples cannot be ignored.

The proposition that detectability increases with the number of embedding operations is heuristic itself, and special cases for counterexamples can be made (cf. the stereo audio signal in the example of Sect. 2.4 or [137]). Nevertheless, distortion minimisation is a good strategy in the absence of knowledge about systematic dependencies (being unaware of dependencies is bad, but it is even worse to create better measurable anomalies that do not exist in natural covers).

Equipped with the overview of Table 3.1, we are now in a position to revisit the last remaining aspect of the relevant related literature.

## 3.4.5 Indeterminacy in the Cover

A necessary requirement for secure steganographic communication is the existence of a public channel, on which sending indeterministic covers is 'plausible'. This finding can be tracked back to the earliest publications on digital steganography; for example, Anderson writes in 1996:

> A purist might conclude that the only circumstance in which she can be certain that [warden] Willy cannot detect her messages is when she uses a subliminal channel in the sense of Simmons; that is, a channel in which she chooses some random bits (as in an ElGamal digital signature) and these bits can be recovered by the message recipient.[4]

A weaker form (because a specific detector is assumed) by Franz, Jerichow, Möller, Pfitzmann and Stierand can be found in the same volume:

> Each stegoanalytical method must work with this nondeterminism. That means it must include gaps. Thus each such stegoanalytical method must accept constellations of bits which cannot be identified as modified. On condition that this stegoanalytical method will be known, these gaps can be used in our algorithms to change these bits anyway. [63]

A proof appeared in [258], which is valid despite the confusion of secrecy and security in the information-theoretic arguments of this paper (cf. footnote 20, p. 98). This is so because the authors prove by contradiction and show that purely deterministic covers break the secrecy of the message. This implies breaking steganographic security. So, indeterministic (from the perspective of the steganalyst) covers are required for secure steganography.

It is further convenient to think of covers as being composed of an indeterministic and a deterministic part. While the former is necessary for steganographic security, the latter is indispensable to ensure plausibility: completely indeterministic covers are indistinguishable from random numbers. If these were plausible, then steganography could be reduced to cryptography.

Disregarding the difficulty of separating the parts exactly, this conceptual distinction is already useful for two reasons.

1. It emphasises that parts of the cover are (better) predictable, over other that are not (worse). This corresponds to our notion of heterogeneity. Steganographic changes should be limited to the indeterministic part of the cover [63, among others]. (The formulations in brackets refer to imperfect separations of indeterministic and deterministic parts.)

2. It helps to formulate conditional cover models, such as the ones proposed by Sallee [206, 207] in his approach to model-based steganography. Note that the separation of what is considered as indeterminisitic or deterministic is part of the model (and as such fallible).

To show that the separation of covers into indeterministic and deterministic parts is fully compatible with our proposed theory, we have to distinguish between two possible sources of indeterminacy.

- **Conditional indeterminacy** exists because the true relation is too complex to make predictions. So, unknown or not well understood mechanisms are substituted by random variables. This applies to all **empirical covers**, the cover generating process of which we believe is incognisable and thus can never be fully understood.
  It is evident that the separation of conditional indeterministic and deterministic parts depends on knowledge about the true relation, which can be expressed in form of hypotheses. Hence, this fits exactly into our notion of cover models. More precisely, we have an indirect cover model with two sub-hypotheses: one for the separation (which can be expressed as a function Proj and, as a special case, should be invertible in order for us to construct efficient embedding functions) and another for the distribution assumption of the so-identified indeterministic part. The distribution function can be conditional on the deterministic part.
- **Unconditional indeterminacy** is introduced by the definition of the channel in **artificial covers**. In this case, the process generating the indeterministic output is fully known to both steganographer and steganalyst. Merely the internal state (i.e., the realisations) are hidden from the steganalyst.[25] The entropy of the indeterminacy can be exploited for secure steganographic communication.

The cover composition approach allows us to combine artificial and empirical covers in a single steganographic system. For example, the weaker form of paradigm II simulates an indeterministic transformation process on empirical covers (Sect. 2.4). Here, the transformation process can be indeterministic *by definition*, similarly to artificial covers, and contributes the indeterministic part of the cover. The empirical cover can safely be used as the deterministic part. It is important to note that the secure capacity of this construction is bounded by the entropy of the artificial part. All modifications to the empirical (i.e., deterministic) part risk detection by a steganalyst with a superior cover model. A refined concept could also separate all three parts, conditionally indeterministic, unconditionally indeterministic and deterministic. But this does not alter the achievable security guarantees.

Two possible caveats of this approach are worth mentioning. First, if the transformation process is merely a partial information-reduction operation

---

[25] If pseudorandom numbers are used in actual implementations, the security guarantees are weakened to the class of complexity-theoretic security.

(e.g., lossy compression), but does not introduce new indeterminacy, then the security cannot be better than in the heuristic class. This is so because we can never rule out the (unlikely) possibility that the steganalyst knows a cover model, which helps predict the original information from the retained cover through higher-order dependencies. The second caveat occurs if the transformation process itself is a model of an empirical process, such as an approximation for sensor noise typically introduced by real sensors [67, 249]. In this case, no security guarantees are possible as the indeterminacy in the process model is conditional (lack of better knowledge), and better approximations of the true process can shift the border between what is deterministic and indeterministic from the steganalyst's perspective.

Despite these caveats, cover transformation processes may still be advantageous as they are easier to grasp and model (knowing that some interdependencies are ignored) than general cover generating processes. Although this does not lead to a higher class of security guarantees, it may defeat more adversaries within the heuristic class. It is reasonable to assume that it is more difficult for a steganalyst to improve upon a relatively well-understood process model than to find just another disregarded dimension in general cover models for empirical covers. Independent of the caveats, another disadvantage of this approach is that covers obviously processed with transformations that are known to be useful in steganography (e.g., adding noise, JPEG double-compression) may in certain environments be less plausible than more diverse covers.

In steganography, as in many other areas, there is no free lunch: ostensible advantages in achievable security have to be paid for with limited capacity, higher embedding complexity, or tweaks with the plausibility heuristic.

## 3.5 Instances of Cover Models for Heterogeneous Sources

After a review of the state of the art in Chapter 2, and the development of a theory of cover models in the previous sections of this chapter, we are now in a position to introduce the selected problems for the specific part of this book. Chapters 4 to 7 each focus on a specific research question in statistical steganalysis of empirical covers. Three of them have in common that the specific advances can be interpreted as instances of the mixture cover model, and the fourth illustrates particularly well the general theory of cover models. Broadly speaking, this means that the detection performance of the proposed steganalysis method benefits greatly by conditional cover models that reduce the heterogeneity of the empirical cover distribution. The validity of all proposed methods is backed with evidence from large sets of real covers.

The most obvious subject of study for a theory of cover models is the approach to model-based steganography by Sallee [206]. The first actual embedding functions derived from the general approach turned out to be the most secure in a (capacity-adjusted) comparison of transformed domain image steganography with a benchmark universal detector [68] in 2004. In Chapter 4 we present a targeted detector which employs a slightly yet effectively improved cover model that detects MB1 steganography with high accuracy without exploiting higher-order statistics. Clearly, raising this constraint could improve the bottom-line detection performance even further, but this is not the point we want to make. Our analysis is deliberately constrained to the 'model world' defined by MB1. And it is striking to see that even extremely simplified cover models leave gaps large enough to construct reliable detectors.

Chapter 5 turns to probably the best understood problem in steganalysis, namely quantitative detectors of LSB replacement in spatial domain representations of greyscale images. Still, the influence of heterogeneous covers has been largely disregarded. Therefore, we present a methodology to identify cover properties that influence detection performance. All properties can be regarded as proxies for realisations of the unobservable random variable in the mixture framework.

Weighted stego image steganalysis (WS, Sect. 2.10.4) is the only quantitative detector for LSB replacement in spatial domain images which employs an explicit cover model. In Chapter 6, we first present enhancements to this cover model that result in substantial performance gains. Then we show how the detection performance can be improved further by incorporating specific cover models for different image origins. In particular, we focus on JPEG pre-compression and propose a new conditional cover model in the mixture framework. The measured performance gains for JPEG pre-compressed are unmatched by other quantitative detectors.

To complement this work with specific results for other cover types than images, a method to estimate the encoder implementation of MP3 audio files as an unobservable random variable in the mixture cover model is presented in Chapter 7. This method allows us to adjust an existing targeted detector [234] to reduce the false alarm rate substantially.

## 3.6 Summary

This chapter has built a formal framework for studying cover models in steganography and steganalysis consistent with the notation and conventions introduced in Chapter 2. although many ideas and concepts for the theoretical underpinnings are adapted from the literature, we believe that the following aspects in this chapter are novel in the steganography community:

- reflection on the plausibility heuristic, and discussion of its consequences;
- explicit framing of the empirical nature of steganography and steganalysis research, and definition of cover models as hypotheses, thereby establishing a link to modern epistemology;

- emphasis of observability bounds for empirical channels and discussion of sampling unit as capacity constraint for provably secure steganography;
- classification of approaches to secure steganography by cover and adversary assumptions;
- idea of mixture distributions to model heterogeneity in empirical covers.

The most prominent open questions concern the relation of observability and computational bounds, and secure capacity limits for the different security classes.

# Part II
# Specific Advances in Steganalysis

# Chapter 4
# Detection of Model-Based Steganography with First-Order Statistics

The research for this chapter was completed in spring 2004 [23]. Here we provide a revised and updated presentation of the original results. Since, a reproduction of the results and a fusion of the original detector with higher-order statistics has been published [228]. These recent advances can be seen as yet another iteration in the search for ever better cover models and should be considered as relevant starting points for future work. However, for the purpose of this book, it is useful to start with the foundations of model-based steganography and its targeted detection methods to show how it fits into our theory of cover models.

This chapter is structured as follows. The next section revisits the ideas of Phil Sallee's approach to model-based steganography [206], using our terminology, and discusses links to (independent) earlier work by Wayner [232], Franz et al. [63] and Zöllner et al. [258]. The actual embedding function of MB1, a specific model-based algorithm for JPEG covers, is described in Sect. 4.2 before we present our proposed targeted detector in Sect. 4.3 and provide experimental evidence for its effectiveness in Sect. 4.4. We discuss limitations of our method and directions for possible and already-realised improved detectors in Sect. 4.5.

## 4.1 Fundamentals of Model-Based Steganography

Model-based steganography is an interesting approach because, for the first time, it included a theoretical underpinning which can be linked to previous information-theoretic approaches to secure steganography. Whereas Sallee [206] developed his concept of model-based steganography partly independently of this literature, we take the opportunity here to provide an alternative presentation of his ideas, thereby acknowledging links to all relevant prior art.

There have been several attempts to formalise the security of stegano-graphic systems with the tools and methods of information theory. The most closely related work to model-based steganography is the paper by Zöllner et al. [258], who postulate the concept of an indeterministic em-bedding function as a necessary requirement for secure steganography (see Sect. 3.4.5). This concept implies that the cover, represented as random vec-tor $\boldsymbol{X} = \boldsymbol{X}_{\mathrm{det}} || \boldsymbol{X}_{\mathrm{indet}}$, can be decomposed into a deterministic part $\boldsymbol{X}_{\mathrm{det}}$ and an indeterministic part $\boldsymbol{X}_{\mathrm{indet}}$. (Informally, this decomposition has al-ready been mentioned by authors of the same research group in [63], called 'nondeterminism' there).

It is conservative to assume that a steganalyst has knowledge about the deterministic part of a specific cover. This knowledge can be of any level of detail, ranging from general knowledge of typical cover statistics to specific (approximations of) realisations of $\boldsymbol{x}_{\mathrm{det}}^{(0)}$ of an actual cover object. Such knowledge can be obtained, for example, by comparing a digitised natural image to the depicted scene or alternative photographs thereof [65, 193]. As a result, any steganographic modification in the deterministic part $\boldsymbol{x}_{\mathrm{det}}^{(0)}$ of a cover is potentially dangerous and must be avoided. For ideal decomposi-tions of $\boldsymbol{X} = \boldsymbol{X}_{\mathrm{det}} || \boldsymbol{X}_{\mathrm{indet}}$, however, the indeterministic part is assumed to be independent[1] and identically distributed (IID) random 'noise'. This noise may originate from random quantisation errors in an analog-to-digital con-verter or from temporal fluctuations in sensor elements. It is assumed that the steganalyst has no knowledge about the actual realisation $\boldsymbol{x}_{\mathrm{indet}}^{(0)}$ of a specific cover, though metainformation such as the proportion and distribu-tion function of the noise may be known. If this assumption holds, it is secure to replace the realisation $\boldsymbol{x}_{\mathrm{indet}}^{(0)}$ with a similarly distributed secret message $\boldsymbol{x}_{\mathrm{indet}}^{(m)}$ to compose a stego object $\boldsymbol{x}^{(m)} = \boldsymbol{x}_{\mathrm{det}}^{(0)} || \boldsymbol{x}_{\mathrm{indet}}^{(m)}$.

Although this approach sounds intriguingly simple in theory, its practical application suffers from the problem of separating $\boldsymbol{X}_{\mathrm{indet}}$ from $\boldsymbol{X}_{\mathrm{det}}$. The decomposition is complicated not only by the varying qualitative assumptions about what information a steganalyst can gain about the cover (this could be worked around by making a conservative assumption, i.e., using a strong adversary model), but also by the impossibility of considering all possible dependencies between elements of empirical covers $\boldsymbol{X}^{(0)}$. Knowledge of *all* interdependencies is required to ensure that $\boldsymbol{X}_{\mathrm{indet}}$ is actually IID after the decomposition. So, as we argue in Sect. 3.4.5 under the heading 'conditional indeterminacy', any practical decomposition of $\boldsymbol{X}_{\mathrm{det}}$ and $\boldsymbol{X}_{\mathrm{indet}}$ implies in fact a cover model.

For example, the (admittedly naïve) decomposition underlying LSB re-placement is that LSBs of $\boldsymbol{x}^{(0)}$ are interpreted as $\boldsymbol{X}_{\mathrm{indet}}^{(0)}$, whereas all re-maining bits belong to $\boldsymbol{x}_{\mathrm{det}}^{(0)}$. LSB replacement is in fact very vulnerable,

---

[1] Here, the attribute 'independent' comprises independence jointly both from reali-sations of other elements in $\boldsymbol{X}_{\mathrm{indet}}$ and from all elements in $\boldsymbol{X}_{\mathrm{det}}$. More formally, $\mathsf{Prob}(x_{\mathrm{indet}\,i} = y) = \mathsf{Prob}\left(x_{\mathrm{indet}\,i} = y | x_{\mathrm{indet}\,j_1}, \ldots, x_{\mathrm{indet}\,j_k}, x_{\mathrm{det}\,\iota_1}, \ldots, x_{\mathrm{det}\,\iota_l}\right) \ \forall i, \ \forall \boldsymbol{j} : \boldsymbol{j} \in ([1, n_{\mathrm{indet}}] \setminus \{i\})^k, \ \forall \boldsymbol{\iota} : \boldsymbol{\iota} \in [1, n_{\mathrm{det}}]^l$.

because this decomposition violates the security requirement that $\boldsymbol{X}_{\text{indet}}^{(0)}$ should be independent of $\boldsymbol{X}_{\text{det}}^{(0)}$ (cf. Sect. 2.10).

Sallee's [206] proposal of a model-based approach to steganography can be interpreted as an evolutionary combination of the decomposition idea and Wayner's [232] mimic functions (cf. p. 45 and Fig. 2.14) coupled with strong implications for the design of steganographic algorithms. In contrast to earlier theoretical work, model-based steganography does not make the unrealistic assumption that $\boldsymbol{X}_{\text{indet}}$ is IID. Instead, it is proposed to find suitable analytical models for the distribution of $\boldsymbol{X}_{\text{indet}}$ conditional on $\boldsymbol{X}_{\text{det}}$. More precisely, a general model (say, parametric distribution function) is fitted to the actual realisation $\boldsymbol{x}_{\text{det}}$ of a specific cover, which leads to a cover-specific model. The purpose of this model is to determine the conditional distributions $\mathsf{Prob}(\boldsymbol{X}_{\text{indet}}|\boldsymbol{X}_{\text{det}} = \boldsymbol{x}_{\text{det}})$. Then, arithmetic decompression known from source (de)coding is employed to skew the distribution of uniform message bits $\boldsymbol{m}$ to the target distribution of $\boldsymbol{X}_{\text{indet}}$ predicted by the model. $\boldsymbol{x}_{\text{indet}}^{(0)}$ can be replaced by $\boldsymbol{x}_{\text{indet}}^{(\boldsymbol{m})}$, which has identical marginal statistics and contains the steganographic semantic of the secret message. Figure 4.1 shows a block diagram of the general model-based embedding process.

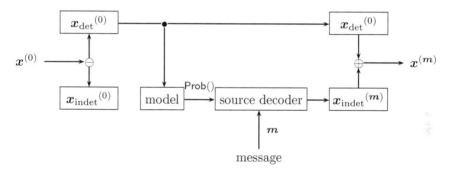

Fig. 4.1: Block diagram of model-based steganography

In addition to these general considerations, the initial work on model-based steganography came up with a proposal for a concrete embedding function for DCT coefficients of JPEG cover. The remainder of this chapter points out weaknesses of this concrete model, which allow a steganalyst to detect stego objects reliably.

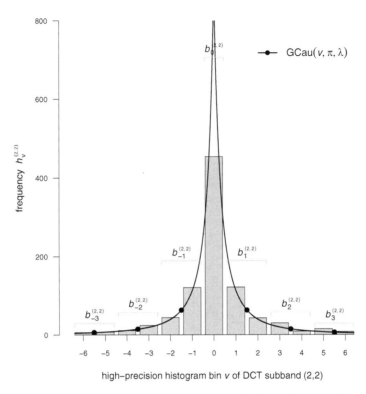

Fig. 4.2: JPEG DCT histogram of subband $(2, 2)$ with generalised Cauchy model superimposed. Fitted parameters are $\pi = 2.38$ and $\lambda = 0.87$

## 4.2 MB1: An Embedding Function for JPEG Covers

In this section, we briefly explain the embedding function MB1 for JPEG images, which was proposed in [206] as an example algorithm for model-based steganography. Like most embedding functions for JPEG covers, MB1 embeds by modifying nonzero values of quantised coefficients of all AC DCT subbands. This ensures that hidden message bits can always be extracted from the resulting JPEG file, as all subsequent compression steps are completely reversible (see Sect. 2.6.4.1 for JPEG compression details).

Since all coefficients in the DCT domain are decorrelated, it is generally believed to be difficult for a steganalyst to (approximately) predict individual coefficient values of the cover from other subband coefficients in the stego object. Therefore, the common approach for targeted detectors of JPEG steganography is based on (subband-specific) histograms aggregated over all blocks in an image, e.g., [18, 78, 152, 153, 238, 243, 244, 248, 252], sometimes taking into account higher-order statistics indirectly via calibrated histograms (see Sect. 2.9.1). To defeat these detectors, MB1 has been designed to approximately preserve all marginal distributions of individual DCT subbands. Figure 4.2 depicts an example histogram of a selected AC subband $(2, 2)$. The histogram shape is typical for all AC subbands. Only the DC subband $(1, 1)$ has a different histogram and is therefore excluded from embedding (as in many other embedding functions for JPEG coefficients). In the terminology of model-based steganography, subband $(1, 1)$ belongs entirely to $\boldsymbol{X}_{\text{det}}$.

In this chapter, let $h_v^{(i,j)}$ be the number of quantised DCT subband $(i, j)$ coefficients equal to value $v$,

$$h_v^{(i,j)} = \sum_{\iota=1}^{k} \delta_{v, y_{i+8j,\iota}}, \tag{4.1}$$

where elements $y$ of matrix $\boldsymbol{y}_{\boxplus}^*$ are obtained as defined in Eq. (2.7). Hence, vector $\boldsymbol{h}^{(i,j)}$ is the histogram[2] of DCT subband $(i, j)$. More precisely, we refer to $h_v^{(i,j)}$ as the $v$th *high-precision* histogram bin. By contrast, *low-precision* histogram bins $b_u^{(i,j)}$ contain two or more adjacent high-precision bins. Without loss of generality, our analysis is limited to the case where each low-precision bin $b_u^{(i,j)}$, $u \neq 0$, contains exactly two high-precision bins. Hence,

$$b_u^{(i,j)} = \begin{cases} h_{2u+1}^{(i,j)} + h_{2u}^{(i,j)} & \text{for } u < 0 \\ h_0^{(i,j)} & \text{for } u = 0 \\ h_{2u-1}^{(i,j)} + h_{2u}^{(i,j)} & \text{for } u > 0. \end{cases} \tag{4.2}$$

To avoid further case differentiation in the notation, we write further equations only for $u > 0$. Similar relations hold for $u < 0$ due to symmetry.

Regarding the decomposition of $\boldsymbol{X}_{\text{det}}$ and $\boldsymbol{X}_{\text{indet}}$, the MB1 algorithm is built on the assumption that the size of low-precision bins $b^{(i,j)}$ belongs to $\boldsymbol{X}_{\text{det}}$, whereas the exact distribution of occurrences within the low-precision bins (i.e., the corresponding high-precision bins $h^{(i,j)}$) is part of $\boldsymbol{X}_{\text{indet}}$.

The embedding function alters quantised AC DCT coefficients so that

1. the new coefficient values stay in their original low-precision bin, and
2. the conditional distribution of $h_{2u-1}^{(i,j)}$ and $h_{2u}^{(i,j)}$ for a given $b_u^{(i,j)}$ conforms to a model of $\boldsymbol{h}^{(i,j)}$ as a function of $(i, j)$.

Effectively, mod-2 replacement is employed as an embedding operation with a mimic function to adjust the target distribution to the model (cf. Sect. 2.7.3). The mimic function takes as input a binary target distribution $\left( \tilde{h}_{2u-1}^{(i,j)}, \tilde{h}_{2u}^{(i,j)} \right)$

---

[2] We slightly extend the vector notation to negative indices $h_v$ with $v \in \mathbb{Z}$.

that depends on both the low-precision bin index $u$ and the subband $(i, j)$. Sallee [206] has proposed a discretised variant of a generalised Cauchy distribution to model the distribution of AC DCT coefficients. The density function is given as

$$\mathsf{PDF}_{\mathrm{GC}}(v; \pi^{(i,j)}, \lambda^{(i,j)}) = \frac{\pi^{(i,j)} - 1}{2\lambda^{(i,j)}} \left( \left| \frac{v}{\lambda^{(i,j)}} \right| + 1 \right)^{-\pi^{(i,j)}}, \qquad (4.3)$$

where parameters $\pi^{(i,j)}$ (shape) and $\lambda^{(i,j)}$ (scale) are fitted with maximum likelihood (ML) for each individual AC DCT subband low-precision histogram,

$$\left( \hat{\pi}^{(i,j)}, \hat{\lambda}^{(i,j)} \right) = \arg\max_{\pi, \lambda} \prod_u \left( \int_{2u-3/2}^{2u+1/2} \mathsf{PDF}_{\mathrm{GC}}(x, \pi, \lambda) \, dx \right)^{b_u^{(i,j)}} \qquad (4.4)$$

$$= \arg\min_{\pi, \lambda} - \sum_u b_u^{(i,j)} \log \int_{2u-3/2}^{2u+1/2} \mathsf{PDF}_{\mathrm{GC}}(x, \pi, \lambda) \, dx. \ (4.5)$$

The generalised Cauchy distribution is symmetric around its fixed mode at 0. Therefore, the location parameter in the standard Cauchy distribution function turns into a shape parameter in this special 'generalised' form. Sallee [206] motivates the choice of this distribution with the apparently better fit than other sharp symmetric distributions, such as the generalised Laplacian, particularly in the tails of the AC DCT histograms (compare Fig. 4.2).

As the overall size of low-precision bins must not be altered, the target distribution of the mimic function is set to match the slope of the high-precision bins within a low-precision bin according to the proportion

$$\frac{\tilde{h}_{2u}^{(i,j)}}{\tilde{h}_{2u-1}^{(i,j)}} = \frac{\mathsf{PDF}_{\mathrm{GC}}(2u; \hat{\pi}^{(i,j)}, \hat{\lambda}^{(i,j)})}{\mathsf{PDF}_{\mathrm{GC}}(2u - 1; \hat{\pi}^{(i,j)}, \hat{\lambda}^{(i,j)})}. \qquad (4.6)$$

It is important to fit the model to low-precision histograms. This ensures that the recipient can recover the model parameters exactly and so obtain the distributions to invert the mimic function.

MB1 was the first practical stego system to employ mimic functions via arithmetic decoding. This leads to embedding efficiencies[3] between 2.06 and 2.16 bits per change for test images compressed with JPEG quality $q = 0.8$ [206]. Comparable algorithms achieve values between 1.0 and 2.0 (OutGuess), 2.0 (JSteg), or slightly below 2.0 (F5).[4] Also, in terms of capacity, MB1 performs on the upper end of the range. MB1 achieves values of just under 14%, which is slightly better than F5 and JSteg (about 13% and 12%, respectively), and clearly above OutGuess (below 7%) [23].

---

[3] See Eq. (2.14); the number of changes appears in the denominator.

[4] F5 achieves higher efficiencies only for embedding rates $p < 2/3$ due to syndrome coding (see Sect. 2.8.2.1).

The developers of MB1 acknowledge that the simple model, being explicitly designed as proof of concept, does not include higher-order statistics and therefore, in principle, has to be considered as vulnerable to detectors which evaluate higher-order statistics. However, Sallee claims it to be "resistant to first order statistical attacks" [206, p. 166].

First-order statistics are all measures describing data regardless of the interdependencies between samples. In other words, first order means invariance to arbitrary permutations of samples. This property applies to histograms and all quantities that can be derived from them (such as moment statistics). By contrast, higher-order statistics consider the relation between samples or their position in the data vector. They include, for example, correlation measures between adjacent pixels in an image.[5]

Our 2004 paper [23] was the first to propose a targeted detector for MB1, although the algorithm has previously been reported to be detectable with Fridrich's [68] universal detector. Her detector exploits higher-order features, such as blockiness measures in the spatial domain and co-occurrence matrices of DCT coefficient values (cf. Table 2.2). Nevertheless, the model-based methods MB1 and MB2 were the most secure ones among the tested algorithms. MB2 [207] complements MB1 with an additional iterative correction procedure to defeat revealing blockiness artefacts [230] at the cost of capacity. Unfortunately, this correction itself leaves even more detectable traces in the image [228].

While researchers' attention was largely focused on the direction of refined higher-order detectors, it is somewhat surprising that MB1 steganography turned out to be also vulnerable from the believed safe side. In the following section, we describe the detection method, which is completely based on first-order statistics.

## 4.3 Detection Method

The core idea of the proposed attack can be summarised as follows: although Sallee's distribution model generally fits the DCT histograms well, there exist outlier bins in natural images. After embedding, these nonconforming bins are adjusted to the slope of the density function of the model distribution. Awareness of the existence of outliers thus constitutes a 'better' cover model in the sense of our theory of Chapter 3. This model advantage can be transformed into a targeted detector.

The construction of our detector can be structured into two steps:

---

[5] Obviously, the definition of first-order statistics depends on the domain: DCT histograms are first-order statistics of the DCT domain but are certainly not in the spatial domain, and vice versa.

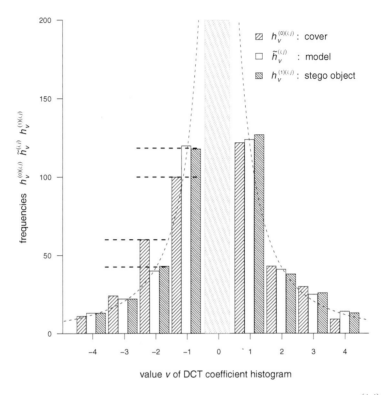

Fig. 4.3: Example DCT histogram with nonconforming bin $b_{-1}^{(i,j)}$

1. A test discriminates nonconforming bins from conforming ones. This test is applied to all low-precision bins of a suspect JPEG image independently.

2. The number of positive test results (i.e., nonconfoming bins) is compared to an empirical threshold for natural covers.

If a suspect image contains fewer nonconforming bins than clean covers usually have, it is likely that the histograms were smoothed by the MB1 embedding function, and thus the image is flagged as a stego object.

Figure 4.3 illustrates a DCT histogram with a typical outlier in the low-precision bin $b_{-1}^{(i,j)}$. Note that the bar of $b_0^{(i,j)}$ is not shown to allow for suitable scaling of the vertical axis. The corresponding coefficients are not used for embedding and therefore are not relevant in this figure. It is clearly visible that the original cover frequencies $h_{-1}^{(0)(i,j)}$ and $h_{-2}^{(0)(i,j)}$ (left bars of the triples) differ from target distributions $\tilde{h}_{-1}^{(i,j)}$ and $\tilde{h}_{-2}^{(i,j)}$ (middle bars). The observed frequencies of the stego object $h_{-1}^{(1)(i,j)}$ and $h_{-2}^{(1)(i,j)}$ (right bars) can be approximated by their expected value via Eq. (4.6) when the output of the mimic function is assumed to follow independent Bernoulli trials.[6] The figure shows that the stego histogram closely fits the target distribution in all bins, whereas some notable deviations (i.e., outliers wrt the model) are observable between the cover histogram and the target distribution.

This mismatch with the cover model can be measured by a contingency test between the observed frequencies and the expected target distribution of both high-precision bins represented in one low-precision bin. To calculate the target distribution, we recover the parameters $\hat{\pi}^{(i,j)}$ and $\hat{\lambda}^{(i,j)}$ similarly to the extraction function. This always leads to identical values as in the embedding function because frequencies of the low-precision histograms are not altered. A contingency table of the form of Table 4.1 is set up to calculate Pearsons's chi-squared ($\chi^2$) statistic. The decision whether or not individual low-precision bins conform to the model is based on a critical value $\chi_{\text{crit}}^2$. This forms a statistical hypothesis test for the null hypothesis that the expected and observed high-precision frequencies are drawn from the same distribution. The test rejects the null hypothesis for nonconforming bins if $\chi^2 > \chi_{\text{crit}}^2$.[7]

Table 4.1: Contingency test for nonconforming low-precision bins

|  | high-precision bin | | |
|---|---|---|---|
|  | left | right | $\Sigma$ |
| observed frequencies | $h_{2u-1}^{(i,j)}$ | $h_{2u}^{(i,j)}$ | $b_u^{(i,j)}$ |
| expected frequencies | $\tilde{h}_{2u-1}^{(i,j)}$ | $\tilde{h}_{2u}^{(i,j)}$ | $b_u^{(i,j)}$ |

To explore the typical share of nonconforming bins in JPEGs generated from natural images, contingency tests were run on the low-precision bins $b_1^{(i,j)}$ and $b_{-1}^{(i,j)}$ for 63 DCT modes of a set of 100 images (altogether 126 tests per image). These images were randomly drawn from a large number of

---

[6] This estimate is consistent even if the arithmetic decoding on average fits the target distribution tighter than does the binomial asymptote of the repeated Bernoulli trials. The histogram preservation method of [62], for instance, would match the target distribution exactly, however at the cost of capacity being bound to the min-entropy rather than the Shannon entropy of the pair of high-precision bins.

[7] The relation to the false acceptance rate for individual hypothesis tests is given via the quantile function of the chi-squared distribution with one degree of freedom [194]; but the interpretation of this metric is not very meaningful in the context of steganalysis. The same applies to the so-called 'chi-squared detector' for simple LSB replacement, where the resulting probability measure should not be interpreted as "roughly the probability of embedding" [238, p. 72].

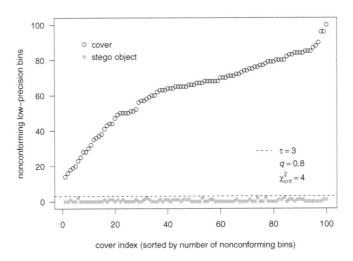

Fig. 4.4: Share of nonconforming bins in natural JPEG images and full capacity MB1 stego objects. Results from 126 contingency tests per image

digital photographs of size $800 \times 600$ pixels and a JPEG quality parameter of $q = 0.8$. Figure 4.4 contrasts the results to 100 full capacity MB1 stego objects created from the same cover images. It is obvious from the figure that a threshold of $\tau = 3$ can reliably discriminate between the two sets.

Two more details of the contingency test are worth mentioning. First, as the test is known to be unreliable (prone to false nonconformities) for small numbers, tables with a minimum frequency below 3 are excluded from the evaluation. Second, the reliability of the test depends on the number of nonzero DCT coefficients. Since this number varies both with the image size and the quantisation factors derived from $q$, the critical value $\chi^2_{\text{crit}}$ can be adjusted to the above mentioned parameters to reach an optimal trade-off of missing rate and false positive rate.

## 4.4 Experimental Validation

The performance of the proposed detection method has been assessed on image set D (cf. Appendix A), which contains about 300 images taken with one of the author's digital camera, an outdated Sony Cybershot DSC-F55E with 2.1 megapixels. Unfortunately, this camera does not support to store images in uncompressed format, so we had to resort to JPEG pre-compressed images. Although the 'simulation' of uncompressed covers from JPEGs is generally difficult, we followed the common approach in the literature (at that time) to reduce potentially unwanted influences or atypical artefacts due to this previous JPEG compression by downscaling all images to a size of $800 \times 600$ pixels [119]. The resulting images have been re-compressed to JPEG with six different quality settings, $q = 0.4, 0.5, \ldots, 0.9$. All experiments are based on greyscale images: chrominance components of colour images have been disregarded. The environment to implement the proposed methods and conduct statistical analyses is the open source *R Project for Statistical Computing* [110, 199], which proved to be a reliable and extensible tool for all data-related work conducted in the course of the preparation of this book.

To generate comparable histogram sets of clean covers and stego objects, all 63 AC DCT histograms were extracted from the test images. The cover histograms were transformed to stego object histograms by simulation: the distribution of the high-precision bins within each low-precision bin was drawn from a binomial distribution with parameters of the target distribution obtained from a fitted generalised Cauchy model,

$$h_{2u-1}^{(1)(i,j)} = r_u^{(i,j)} \qquad \text{with} \qquad R_u^{(i,j)} \sim \mathcal{B}\left(b_u^{(i,j)}, \frac{\tilde{h}_{2u-1}^{(i,j)}}{b_u^{(i,j)}}\right), \text{ and} \qquad (4.7)$$

$$h_{2u}^{(1)(i,j)} = b_u^{(i,j)} - h_{2u-1}^{(1)(i,j)}. \qquad (4.8)$$

Furthermore, it is obvious that shorter secret messages imply less change to the histogram, and thus lower detectability. Since MB1 does not use any coding techniques to improve embedding efficiency, it is safe to assume that the impact on the histograms decreases proportionally with the length of the secret message.[8] To assess the influence of message length, we also varied $p$ in ten steps from full capacity down to 10% for all test images and quality factors. This led to a set of 1.2 million stego DCT histograms (equivalent to 18,120 stego objects), which were compared to the corresponding cover sets.

Prior exploration of suitable bins for the contingency test revealed that the bins $b_{-1}^{(i,j)}$ and $b_1^{(i,j)}$ yielded the best results for all DCT subbands. So we decided to ignore all other bins in our detection algorithm. We also ran the proposed detector on a smaller set of independent images, though taken from the same source camera, to determine suitable parameters (for fixed $p = 1$ and $q = 0.8$). We found that all images could be correctly classified with $\chi^2_{\text{crit}} = 6$ and threshold $\tau = 2$. We systematically varied $\chi^2_{\text{crit}}$ to generate ROC curves while keeping $\tau = 2$ constant for all experiments.[9] We

---

[8] This assumption is based on the absence of any metainformation, such as headers, and an asymptotically ideal mimic function. If one of these criteria is not met, our detectability estimates tend to *underestimate* the true detectability. So we would err on the safe side.

[9] The common approach to measure the error rates as a function of $\tau$ (cf. Sect. 2.3.2) is not suitable in this case, because the underlying quantity (number of nonconforming bins) is discrete. Of course, other aggregation formulae than this two-step test-and-count approach are conceivable as well, but remain beyond the scope of this work.

acknowledge that further 'optimisations' might be possible by adjusting $\tau$, but we refrained from doing so to avoid overfitting (which risks compromising the generalisability of our result) on a relatively small database of rather homogeneous test images.

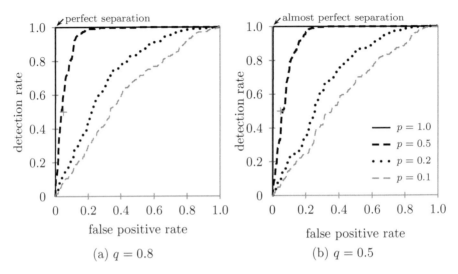

Fig. 4.5: ROC curves of detection performance against MB1 for varying embedding rates $p$ and JPEG qualities $q$; $N = 300$ JPEG covers

Figure 4.5 shows ROC curves of the detection performance for selected embedding rates and JPEG qualities. As the results of Fig. 4.4 already suggest, we also reached good discriminatory power for the test set. It is very well visible that full capacity embedding is perfectly detectable, and the detection power degrades only slowly with decreasing embedding rates. This demonstrates that in fact the MB1 algorithm can be 'broken' (i.e., is detectable) with first-order statistics.

It is common to quantify the information in the ROC curve in a measure of detection reliability, for which we choose normalised AUC (cf. Sect. 2.3.2). The respective measures are printed in Table 4.2. Both the table and the ROC plots also indicate Ker's criterion of 'reliable steganalysis', which corresponds to lower than 50% misses at 5% false positives [118]. This point is marked with a grey cross in Fig. 4.5, and table entries printed in boldface indicate that the criterion is met.

Observe that the detection reliability is barely affected by the JPEG quality. This is surprising at first sight because one would expect that the existence of outliers depends largely on the number of nonzero DCT coefficients, i.e., highly populated histograms for high JPEG qualities should, on average,

Table 4.2: AUC measures of detection performance against MB1

| embedding | avg. message size | JPEG quality $q$ | | | | | |
|---|---|---|---|---|---|---|---|
| rate $p$ | (% of file size) | 0.9 | 0.8 | 0.7 | 0.6 | 0.5 | 0.4 |
| 1.0 | 13.1% | **1.0000** | **1.0000** | **0.9999** | **1.0000** | **1.0000** | **0.9979** |
| 0.9 | 11.8% | **1.0000** | **1.0000** | **0.9997** | **1.0000** | **0.9996** | **0.9963** |
| 0.8 | 10.5% | **0.9989** | **0.9982** | **0.9949** | **0.9970** | **0.9940** | **0.9826** |
| 0.7 | 9.2% | **0.9890** | **0.9850** | **0.9797** | **0.9777** | **0.9740** | **0.9596** |
| 0.6 | 7.9% | **0.9593** | **0.9527** | **0.9440** | **0.9322** | **0.9292** | **0.9202** |
| 0.5 | 6.6% | **0.9012** | **0.8898** | **0.8782** | 0.8615 | 0.8552 | 0.8519 |
| 0.4 | 5.2% | 0.8057 | 0.7906 | 0.7796 | 0.7624 | 0.7509 | 0.7516 |
| 0.3 | 3.9% | 0.6576 | 0.6476 | 0.6457 | 0.6185 | 0.6063 | 0.6180 |
| 0.2 | 2.6% | 0.4568 | 0.4549 | 0.4583 | 0.4295 | 0.4214 | 0.4362 |
| 0.1 | 1.3% | 0.2289 | 0.2294 | 0.2357 | 0.2160 | 0.2133 | 0.2252 |

The ROC curves for values printed boldface meet a reliability criterion of more than
50% detection rate with less than 5% false positives; $N = 300$ JPEG pre-compressed
covers.

be less prone to outliers. However, a further analysis of the contribution of
nonconfoming bins per DCT subband reveals that for each quality, there
exist 'borderline subbands' which hold few (but not too few to render the
contingency test unreliable) nonzero DCT coefficients. As can be seen in Figure 4.6, with increasing JPEG quality, the 'dangerous' subbands move gradually towards the higher frequencies. Hence, while the generalised Cauchy
model seems appropriate to cope with different quantisation factors through
its scale parameter $\lambda$, it provides no mechanism to reflect the high variance
of a binomial distribution when its first parameter (the number of Bernoulli
trials) is small. This seems to be the root cause of nonconforming bins.

## 4.5 Summary and Outlook

This chapter has provided a real-world example to support the hypothetical
evolution of cover models (a) to (c) in Sect. 3.2.3. It has been demonstrated
how an embedding function that has previously been considered as secure can
successfully be detected by a refinement of the steganalyst's cover model. The
case of MB1 is particularly interesting for two reasons:

1. MB1 is a very good case study for our theory of Chapter 3, as it has been
   designed with an explicit cover model.

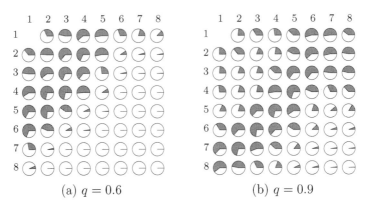

(a) $q = 0.6$                    (b) $q = 0.9$

Fig. 4.6: Distribution of nonconforming bins over subbands for varying JPEG compression qualities $q$

2. The fact that our detector is based on first-order statistics means that the proposed detector does not even need to take advantage of the steganographer's difficulty of coping with high dimensionality of natural covers; it uses the same projection Proj as the embedding function, which has merely been refined by a slightly superior measurement that considers outliers.

The remainder of this section is devoted to three open subjects. First, we point to limitations of the proposed detector and propose future improvements. Then, we discuss possible countermeasures to defeat this simple detector, before we come to more general implications for the design of new and better embedding functions.

### 4.5.1 Limitations and Future Directions for MB1 Steganalysis

A possible extension to this detector can be the anticipation of common image processing steps. As the experiments were run, by today's standards, on a rather small set of test images from a single digital camera, we cannot generalise our results to all kinds of covers. For example, natural images are often subject to image transformation and manipulation. It is likely that some of these algorithms, e.g., blurring, also result in smoother DCT histograms. The challenge in detecting these manipulations in advance and thus reducing the number of false positives, or even in distinguishing between 'white hat' image processing and 'black hat' Cauchy model-based steganography, is subject to further research. Although not focused on the case of MB1 detection, the following chapters proceed in this direction. We investigate more closely how heterogeneity in images can be identified and modelled (Chapter 5) to be ultimately incorporated in conditional cover models of cover-source-specific detectors (Chapter 6 for image and Chapter 7 for audio covers).

## 4.5.2 Possible (Short-Term) Countermeasures

In thinking about possible countermeasures, the ad hoc solution, namely capacity reduction, is as old as digital steganography. But the assumed positive relation between embedding distortion and detectability is just a heuristic. It remains an interesting research question how lower capacity can be optimally transformed into security gains. One option is to employ syndrome coding to reduce the distortion per message bit, possibly combined with wet paper codes to exclude outlier bins. Modest compromises in capacity can also be used to achieve a perfect preservation of first-order statistics as suggested independently in [198] and [219, 220], thereby making the parametric model dispensable.

Refining the model could be another countermeasure. Strict preservation of all regarded cover properties (often combined with even worse distortion of ignored properties) has shown to be pretty ineffective, and also complicates the embedding function unnecessarily. Therefore, we could imagine modelling a certain amount of nonconforming bins, and randomly interspersing the stego object with outliers. But care must be taken: the steganalyst may predict the location of outliers (we have not investigated this so far) and thus gain advantage if the steganographer's model inserts outliers at random or in a predictable fashion.

Our MB1 detector emphasises again the risk of giving up the information superiority of the steganographers to third parties who do not know the secret key. MB1 is successfully detectable, among other things, because the steganalyst is able to recover the model parameters exactly.[10] Hence, as a design principle, new algorithms should consider either making the parameter retrieval dependant on the secret key, or implementing an embedding function which does not require the recipient to know the exact model. Admittedly, this insight was more relevant in 2004 and has become a generally accepted standpoint in the meantime. With the discovery of wet paper codes (cf. 2.8.2.2), the right tools to realise this principle are readily available and in common practice.

---

[10] Note that hiding the exact parameters does not prevent a steganalyst from estimating them approximately. As macroscopic properties of covers are difficult to hide completely, they should never be regarded as a secret.

## 4.5.3 Implications for More Secure Steganography

When reasoning about the implications of the detector presented, it is important to emphasise that this vulnerability of MB1 is rather a problem of the specific cover model of this embedding function than a weakness of the general approach of model-based steganography. Also, the refinement to MB2 [207], which included a new defense against higher-order attacks with blockiness measures [230], was not very successful. As shown in [227], the 'overfitting' of a single criterion for blockiness made it vulnerable to detectors which verify this criterion in comparison to other alternative measures that are not considered in the iterative correction algorithm of MB2.

Despite these specific problems of MB1 and MB2, the model-based approach paved the way in a promising direction towards systematically designed steganographic algorithms. The clear link between information-theoretic considerations and the actual algorithms contributes to structuring the research area and integrates well in our theoretical framework. It is even conceivable to design model-based embedding functions with conditional cover models, such as models that treat empirical covers as a mixture source.

# Chapter 5
# Models of Heterogeneous Covers for Quantitative Steganalysis

Unlike the other chapters in this part of the book, this one does not present a new or improved detector, but contributes a methodology to deal with sources of heterogeneous covers in the context of steganalysis. In general, the detectability of secret messages, and hence the success of steganalysis, depends on many factors, which can be attributed to

1. the embedding function Embed and (implicitly) the embedding domain, or
2. the detection method Detect, or
3. the message $m$ (for uniform random bit strings, mainly its length matters), or finally
4. the cover $x^{(0)}$ (e.g., properties related to source, size, preprocessing arte-facts, etc.).

Of this list, embedding function and detection methods are usually at the centre of interest in academic papers. Since authors have to demonstrate the performance of each new method compared to existing ones in order to ensure publication, the influence of the first two bullet points is generally best understood. However, studies of other influencing factors than message length in conjunction with cover size (i.e., the net embedding rate $p$) are pretty rare. This is particularly unfortunate, as a systematic approach to dealing with heterogeneous covers requires a better understanding of the individual and joint influences of various cover properties. Moreover, occasional evidence in the literature has shown that cover properties can matter a lot [118, 119, 202, among many others]. Therefore, this chapter is devoted to methodological questions in identifying and quantifying the influence of cover properties for existing detectors of one well-understood embedding algorithm, namely LSB replacement. We decided to focus our study on LSB replacement in spatial domain image representations because a wide range of partly analytically well-understood and reasonably accurate quantitative detectors exist for this embedding operation and cover type. Quantitative detectors (cf. Sect. 2.9.3) are particularly suitable for studying the influence of cover properties because they offer great flexibility in defining various performance measures on continuous scales, which at the same time provide a clear link to the decision threshold $\tau$ in binary steganalysis and to the related error rates $\alpha$ and $\beta$ (cf. Sect. 2.3.2).

To structure the presentation in this chapter, we first review common performance metrics for quantitative detectors in Sect. 5.1 and discuss their properties against the backdrop of the distribution of estimates $\hat{p}$ that can be observed in practice. We further refine these metrics by separating two main sources of estimation errors: cover and message. In Sect. 5.2, we propose regression analysis based on parametric distribution models as a tool to evaluate steganalytic performance and its interdependence with explanatory variables. By using macroscopic cover properties as explanatory variables, the method can be applied to identify and quantify the influence of cover properties on detection performance (Sects. 5.2.2 and 5.2.3). This can help us, inter alia, to select conditional cover models that reflect heterogeneity in the distribution of cover properties best.

The presentation here is not intended to be a comprehensive benchmarking exercise of all relevant quantitative LSB detectors, but rather a structured discussion of issues arising in quantitative steganalysis of heterogeneous covers. We refer the reader to our publications [15] and [22], which contain more details on benchmarking aspects. The results reported here are based on unpublished data generated in the same research effort. The main difference with the published result is that in [22], the message lengths instead of the number of flipped LSBs were fixed.

## 5.1 Metrics for Quantitative Steganalysis

Quantitative stego detectors estimate the net embedding rate

$$\hat{p} = \mathsf{Detect}_{\mathsf{Quant}}(\boldsymbol{x}^{(p)}) \tag{5.1}$$

of a suspect object $\boldsymbol{x}^{(p)}$. They can be converted to binary stego detectors (discriminators) by testing the null hypothesis $p = 0$ via $\hat{p}$. The distribution of the estimation error $\hat{p} - p$ determines the achievable error rates $\alpha$ and $\beta$ for the detection of stego objects with given embedding rate $p$.

### 5.1.1 Conventional Metrics

Figure 5.1 depicts the typical charts, by which results of quantitative detectors were reported in the literature (for instance, in [73, 74, 120, 160]). It plots the estimation results for each element of a limited set of 64 test images at various net embedding rates $p$. The two exemplary detectors compared

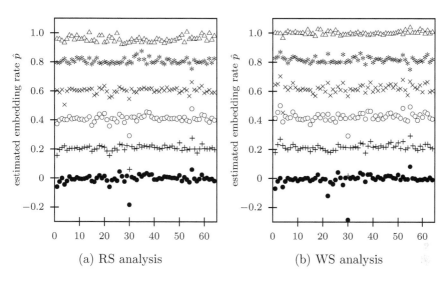

(a) RS analysis                    (b) WS analysis

Fig. 5.1: Estimates of $\hat{p}$ from 64 never-compressed greyscale images of image set B; $p = 0\ (\bullet), p = 0.2\ (+), p = 0.4\ (\circ), p = 0.6\ (\times), p = 0.8\ (*), p = 1\ (\triangle)$

Table 5.1: Conventional performance measures for quantitative detectors

|  | Embedding rate $p$ | | | | | |
|---|---|---|---|---|---|---|
|  | 0.0 | 0.2 | 0.4 | 0.6 | 0.8 | 1.0 |
| **Mean** | | | | | | |
| RS analysis | −0.005 | 0.205 | 0.409 | 0.603 | 0.806 | 0.955 |
| WS analysis | −0.009 | 0.213 | 0.424 | 0.620 | 0.817 | 1.000 |
| **Standard deviation** | | | | | | |
| RS analysis | 0.031 | 0.026 | 0.025 | 0.022 | 0.020 | 0.021 |
| WS analysis | 0.044 | 0.033 | 0.029 | 0.034 | 0.021 | 0.013 |

Parameter estimates from $N = 64$ never-compressed greyscale images.

here are described in Sects. 2.10.1 (RS analysis) and 2.10.4 (WS analysis), respectively.

The most obvious quantitative performance metrics derived from such data are estimates of moment statistics of the data points, such as the mean as a measure of *bias* and standard deviation as a measure of *dispersion* (called *accuracy* in [15]). Table 5.1 shows a typical result tabulation of these measures corresponding to the data depicted in Fig. 5.1. The implicit assumption in reporting such metrics is that the moment statistics do converge (quickly enough, for small $N$). But this might not be the case in the presence of outliers, such as the images with indices 30 and 55 in Fig. 5.1.

A common approach in statistics is to eliminate outliers before conducting statistical inference tests to avoid unwanted influence. However, the high number of outliers observable in quantitative steganalysis makes them the most influential factor for misclassifications of cover and stego objects, particularly at low embedding rates. This justifies treating outliers as genuine, and calls for solutions that accommodate rather than eliminate them. A straight approach is to model the estimation error $\hat{p} - p$ with a fat-tailed distribution. This not only allows for more consistent estimates of summary statistics, but is also crucial for the design of appropriate regression models to compare detectors and identify influential cover properties (cf. Sect. 5.2 below).

### 5.1.2 Improved Metrics Based on a Distribution Model

The observation that embedding rate estimates of RS and SPA are non-Gaussian and leptokurtic (i.e., longer-tailed, which means excess kurtosis above 0) has been first noted by Ker [119]. This early publication framed the finding as a cautionary hint and refrained from giving clear indication on how to deal with the phenomenon. In [15] we reported a good fit of the Cauchy distribution for RS and (standard) WS estimation errors on the basis of likelihood ratio tests against two alternative two-parameter distributions, namely Gaussian and Laplacian. We have refined the model assumptions to a Student $t$ distribution in [22].

The class of Student $t$ distributions, of which the Cauchy distribution is a special case, offers more flexibility in handling long-tailed data and shall be used here. Recall the probability density function (PDF) of the Gaussian normal distribution with location parameter $\mu$ and scale parameter $\sigma > 0$,

$$\mathsf{PDF}_{\mathcal{N}}(u; \mu, \sigma) = \underbrace{\frac{1}{\sigma\sqrt{2\pi}}}_{\text{scaling const.}} \underbrace{e^{-\frac{(u-\mu)^2}{2\sigma^2}}}_{\text{shape}}, \tag{5.2}$$

and compare the density function of the Student $t$ distribution with an additional *degrees of freedom* parameter $\nu > 0$,

$$\mathsf{PDF}_{t}(u; \mu, \sigma, \nu) = \underbrace{\frac{\Gamma\left(\frac{\nu+1}{2}\right)}{\sigma\sqrt{\nu\pi}\,\Gamma\left(\frac{\nu}{2}\right)}}_{\text{scaling constant}} \underbrace{\left(1 + \frac{(u-\mu)^2}{\sigma^2\nu}\right)^{-\left(\frac{\nu+1}{2}\right)}}_{\text{shape}}. \tag{5.3}$$

Here $\Gamma$ denotes the Gamma function and $\pi = \Gamma(\frac{1}{2})^2$ is the radial constant. When $\nu = 1$, the Student $t$ distribution reduces to the Cauchy distribution, and as $\nu \to \infty$, it approaches the Gaussian distribution.

Observe that the argument $u$ appears in the exponent of the 'shape' term of $\mathsf{PDF}_\mathcal{N}$ (Eq. (5.2)), whereas it is inverse polynomial in $\mathsf{PDF}_t$ (Eq. (5.3)). This creates the important difference in asymptotic behaviour at the tails: while the Gaussian cumulative density function (CDF) decays exponentially, i.e., $1 - \mathsf{CDF}_\mathcal{N} \propto e^{-u}$, the probability of outliers of the Student $t$ distribution decreases only with the reciprocal of a polynomial in $u$, that is, $1 - \mathsf{CDF}_t \propto u^{-\nu}$. Parameter $\nu$ controls the fatness of the tails and is also referred to as *tail index* for this reason. An important property of the Student-$t$ distribution is that only moments strictly smaller than $\nu$ are finite. This is why data presumably following a Cauchy distribution ($\nu = 1$) cannot be summarised by empirical mean estimates, and reporting standard deviations is only reasonable if the variance exists, i.e., $\nu > 2$.

Parameters of the Student $t$ distribution can be estimated from data with maximum likelihood (ML),

$$(\hat{\mu}, \hat{\sigma}, \hat{\nu}) = \arg\max_{\mu, \sigma, \nu} \prod_{i=1}^{N} \mathsf{PDF}_t(\hat{p}_i; \mu, \sigma, \nu) \tag{5.4}$$

$$= \arg\min_{\mu, \sigma, \nu} -\sum_{i=1}^{N} \log \mathsf{PDF}_t(\hat{p}_i; \mu, \sigma, \nu), \tag{5.5}$$

for example by using the scoring method by Taylor and Verbyla [224]. This method is a modified Fisher scoring procedure that alternatingly updates scale and location terms to accelerate convergence. For quantitative detectors with Student $t$-like error distributions, parameter estimates from this procedure can be interpreted as performance metrics even in cases where Gaussian moments are inappropriate.

Figure 5.2 provides evidence that estimation errors $\hat{p}$ obtained from running $\mathsf{Detect}_{\mathsf{Quant}}$ on covers ($p = 0$), a quantity called 'initial bias' in [74], can be well modelled with distributions of the Student $t$ family. Results for RS analysis (cf. Sect. 2.10.1), WS analysis (Sect. 2.10.4) and SPA (Sect. 2.10.2) are all shown in two different ways:

1. The **charts on the left** depict histograms of the error distribution (scaled to frequency density) with density functions of fitted Gaussian and Student $t$ distributions superimposed. Observe that the Student $t$ family captures the shape of the error distribution much better than the Gaussian alternative, although these diagrams are suitable to confirm the fit in the centre of the distribution and hide details of the fit on the tails.
2. The **charts on the right** represent the empirical and theoretical cumulative distribution. Plotted on double-log scales, these diagrams are

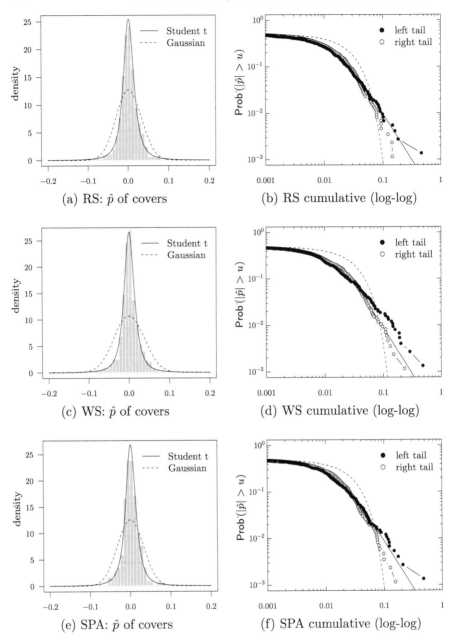

Fig. 5.2: Fitted distributions of estimation error $\hat{p} - p$ for $p = 0$; data from 800 never-compressed greyscale images (set B); location and scale parameters estimated; $\nu = 2$

particularly suitable for inspecting the fit on the tails.[1] Note that negative values on the left tail of the (symmetric) distributions are mapped to the same scale by taking absolute values. Tails can be directly associated with the error rates of the steganalysis decision: The right tail is responsible for false positives whereas a long left tail, for $p > 0$, increases the number of misses. It is clear that the Student $t$ distribution is a much better fit to the observed data than the Gaussian distribution also with regard to the tails. We caution the reader against giving too much significance to the last few data points, because they represent only the highest and lowest few observations in the sample and, like all extreme value statistics, are unstable and prone to the existence of outliers. Nevertheless, a small systematic asymmetry in the tails is common to all detectors: the left tail is somewhat 'fatter' than the right tail. It is up to future research to find out if this asymmetry is specific to the sample or a characteristic of the detectors; and to consider if modelling left and right tails separately is beneficial for any purpose.

Similar charts for two more quantitative steganalysis methods, the least-squares solution to SPA (abbreviated SPA/LSM; Sect. 2.10.3.1) and Triples as a representative of higher-order structural steganalysis (Sect. 2.10.3.3), are reported in Fig. G.1 in Appendix G.

One problem in fitting Student $t$ distributions to data is the difficulty of estimating $\nu$ accurately. As can be seen in Fig. 5.3, the likelihood has a rather gentle slope as $\nu$ varies; this is so because a slight misspecification of $\nu$ can be compensated with adjustments in the scale parameter. Moreover, the apparent maximum at $\nu = 2$ is problematic, because $\nu = 2$ is the critical value, only above which variance is finite. Finite variance is a precondition for many 'traditional' statistical methods, including the central limit theorem (CLT). However, even if the true value of $\nu$ turns out to be slightly above 2, any such asymptotic result will have extremely slow convergence and should not be relied upon in a limited sample of several hundred observations. To be on the safe side, we completely avoid 'traditional' methods and use the Student $t$ distribution model, which can be fitted to data by maximising the likelihood function in spite of undefined moments.

To work around the problem of estimating $\nu$ and to establish a base for comparison of scale parameters between detectors, we decided to fix $\nu = 2$ for all fitted Student $t$ distributions. The numerical results in Table 5.2 support our proceeding: the differences in goodness of fit (measured by log likelihood) between fixed $\nu = 2$ and estimated $\nu$ (following the method in [224]) are rather small, and so are the differences in estimated scale and location parameters. As estimates for $\nu$ range between 1.58 (Triples) and 2.40 (RS), $\nu = 2$ appears a conservative compromise because it is unlikely that the true $\nu > 2$ for all detectors. So comparisons between detectors on the

---

[1] These plots were inspired by diagrams in [29], a textbook on financial mathematics written by physicists.

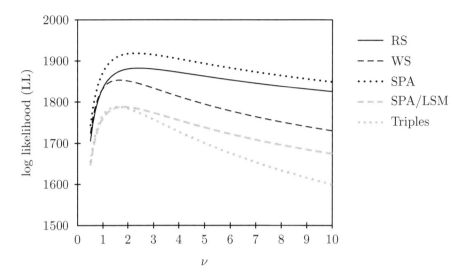

Fig. 5.3: Log-likelihood profiles as a function of $\nu$; fitted to data from $N = 800$ never-compressed greyscale images (set B)

basis of second and higher moment statistics are inappropriate and, as can be seen from the rightmost columns, may produce misleading information. (Most remarkably, Triples appears severely biased and comparatively less accurate using Gaussian moments instead of Student $t$ parameters; the same applies to SPA/LSM, albeit to a lesser extent.)

For sufficiently large samples, nonparametric measures of bias and dispersion offer another alternative to Gaussian moments. Metrics such as median and interquartile ranges (IQRs) are easy to calculate, do not require any distribution assumption, and are immune to outliers. However, as no particular distribution is assumed, they cannot be linked to the error rates $\alpha$ and $\beta$.

Although there is no theoretical justification for the Student $t$ distribution of estimation errors, having a good parametric model for the errors is an important prerequisite for drawing statistical inferences in benchmarking of different detectors and in identifying (and quantifying) the influence of image-specific properties. However, before we can advance to the latter subject, we have to regard the composition of the error distribution $\hat{p} - p$ when real stego objects (i.e., $p > 0$) are presented to $\mathsf{Detect_{Quant}}$.

Table 5.2: Performance metrics based on the Student $t$ assumption for $\hat{p} - p$

| | Student $t$ | | | | | | Gaussian | |
| | $\nu = 2$ fixed | | | $\nu$ estimated | | | moments | |
| Method | location scale [a) b)] | LL | location scale [a) b)] | $\hat{\nu}$ [b)] | LL | $\hat{\mu}$ | $\hat{\sigma}$ |
|---|---|---|---|---|---|---|---|---|
| RS | 0.18** (0.063) 1.39 (0.056) | 1881 | 0.19** (0.066) 1.47 (0.056) | 2.40 (0.243) | 1882 | 0.12 | 3.15 |
| WS | 0.12* (0.061) 1.33 (0.054) | 1851 | 0.11 (0.058) 1.23 (0.053) | 1.66 (0.133) | 1854 | 0.02 | 3.73 |
| SPA | 0.16** (0.060) 1.32 (0.053) | 1918 | 0.17** (0.062) 1.37 (0.053) | 2.29 (0.224) | 1919 | 0.06 | 3.18 |
| SPA/LSM | 0.08 (0.067) 1.47 (0.059) | 1788 | 0.08 (0.066) 1.42 (0.059) | 1.81 (0.153) | 1788 | −0.29 | 4.21 |
| Triples | −0.10 (0.064) 1.41 (0.057) | 1784 | −0.08 (0.061) 1.30 (0.056) | 1.58 (0.124) | 1789 | −1.05 | 6.12 |

Estimates from $N = 800$ never-compressed greyscale images (set B, $p = 0$); unit scaled to percentage points of the net embedding rate; standard errors in brackets; significance levels: *** $\leq 0.001$, ** $\leq 0.01$, * $\leq 0.05$.

[a)] fitted in log of the square; std. errors approximated for untransformed scale

[b)] no significance test computed due to lack of null hypothesis

## 5.1.3 Decomposition of Estimation Errors

An additional source of variation in the estimation error of stego objects ($p > 0$) is the realisation of the embedded secret message. For fixed net embedding rate $0 < p \leq 1$, cover size $n$, and quantitative detector $\mathsf{Detect_{Quant}}$, let

$$\hat{p}_{i,j} = \mathsf{Detect_{Quant}}\left(x_i^{(m_j)}\right), \text{ with } |x_i| = n \text{ and } |m_j| = \lceil np \rceil, \quad (5.6)$$

be the detector output for the $j$th realisation of all possible messages $m \in \mathcal{M}$ of equal length $np$, embedded in the $i$th realisation drawn from the distribution of covers $x^{(0)} \in (\Omega, \mathcal{P}_0)$ of length $n$. In the absence of better knowledge of actual realisations of the cover and message in a specific instance, we have to assume random draws. This suggests that $\hat{p}_{i,j}$ can be written as random variable $\hat{P}$. Its deviation from true $p$ can be decomposed by the source of randomness:

$$\hat{P} - p = Z_{\text{cover}} + Z_{\text{message}}. \quad (5.7)$$

Random variable $Z_{\text{cover}}$ models the part of the estimation error due to the selection of the $i$th cover and is called *between-image error* distribution.

Conversely, $Z_{\text{message}}$ explains the variation of estimates for a fixed image over varying messages and therefore is referred to as *within-image error* distribution [22]. We assert $\mathsf{E}(Z_{\text{message}}) = 0$ and thus assign all bias to the between-image distribution to ensure that this decomposition remains identifiable.

As the focus of our interest is on heterogeneity between covers, we have to search for ways to eliminate or attenuate the influence of $Z_{\text{message}}$ in our measurements. In [22] we have shown that the magnitude of within-images errors is not negligible in some cases and should not be ignored (see also Table 5.4 below).

In fact, theoretical considerations suggest that $Z_{\text{message}}$ can be further decomposed into two components,

$$Z_{\text{message}} = Z_{\text{flips}} + Z_{\text{pos}}. \tag{5.8}$$

This is so because all quantitative detectors of LSB replacement estimate $\hat{p}$ via the expected number of flipped LSBs $\frac{1}{2}np$ (for uniform messages, on average 50% of the embedding positions already contain the correct steganographic semantic; cf. footnote 35 on p. 53). However, $\frac{1}{2}np$ is only the average number of flips and the actual value may fluctuate around this amount according to a binomial distribution, $Z_{\text{flips}} + \frac{1}{2}np \sim \mathcal{B}\left(np, \frac{1}{2}\right)$, depending on both cover and message.[2] The remaining variation per message can be attributed to the distribution of embedding positions over the cover samples, controlled by the key-dependent random permutation. Random correlation of message bits and sample values at the embedding positions materialise in small offsets of $\hat{p}$. In fact, the number of flips and the interleaving path are mutually interdependent, but since we possess a theoretical model only for $Z_{\text{flips}}$, we decided to model the covariation as part of $Z_{\text{pos}}$.

### 5.1.3.1 Shape of the Error Distribution Components

We use an empirical approach to explore the shape and parameters of $Z_{\text{pos}}$ by repeatedly simulating stego objects with exactly $\frac{1}{2}np$ LSBs flipped (hence, $\mathsf{E}(Z_{\text{flips}}^2) = 0$). Detect$_{\text{Quant}}$ is then run on $\iota = 200$ stego objects *per cover* to generate samples from $Z_{\text{pos}}$. We claim that $Z_{\text{pos}}$ is approximately Gaussian and provide evidence in Table 5.3. It reports the percentage of error distributions for which the Shapiro–Wilk test [203], chosen for its good power in discriminating Gaussian distributions from heavy-tailed alternatives, cannot reject the Gaussian null hypothesis at a significance level of 10%. Each cell summarises $\iota = 200$ detection results of each of $N = 800$ images. Cells with an insufficient number of observations (due to detector failures at higher embedding rates in the case of SPA/LSM, and even more so in the case of Triples)

---

[2] If encoding techniques are used to minimise the number of changes (cf. Sect. 2.8.2.1), then the distribution of $Z_{\text{flips}}$ can become a much more complicated function of the gross embedding rate. We do not regard such cases in this work.

Table 5.3: Summary of Shapiro–Wilk tests for normality of $Z_{\mathrm{pos}}$

| | Number of LSB flips as fraction of $n$ | | | | | | | |
|---|---|---|---|---|---|---|---|---|
| | 0.005 | 0.025 | 0.05 | 0.1 | 0.2 | 0.3 | 0.4 | 0.5 |
| Method | ($p$=0.01) | ($p$=0.05) | ($p$=0.1) | ($p$=0.2) | ($p$=0.4) | ($p$=0.6) | ($p$=0.8) | ($p$=1.0) |
| RS | 91.1 | 88.1 | 88.8 | 89.6 | 88.8 | 88.8 | 66.4 | 83.4 |
| WS | 87.0 | 90.1 | 91.1 | 90.6 | 89.0 | 90.0 | 89.4 | 89.0 |
| SPA | 88.9 | 88.9 | 89.0 | 89.9 | 90.8 | 87.2 | 44.0 | 83.0 |
| SPA/LSM | 89.6 | 88.9 | 88.6 | 90.1 | 89.5 | 87.1 | | |
| Triples | 90.6 | 88.8 | 88.4 | 88.8 | | | | |

Percentage of $N = 800$ distributions from never-compressed greyscale images (set B) with $\iota = 200$ data points each, for which the Shapiro–Wilk test cannot reject the null hypothesis of normality at a 10% significance level. If the normality assumption is true, we would expect a pass rate of 90%.

are intentionally left blank. The table confirms that, in a very wide variety of circumstances, $Z_{\mathrm{pos}}$ has a Gaussian distribution. The only obstacle which we cannot explain is the drop at $p = 0.8$ for all detectors but WS. More detailed inspection revealed that the deviation from normality is due to a singular positive skew, which is only observable at this embedding rate (but remains broadly stable when the images are JPEG pre-compressed, cropped or resized to smaller dimensions or if the red channels of the original images are analysed instead of the luminance channels). Our initial suspicion that the singularity may be related to the exclusion of cases when the estimation equation has no real root is difficult to square with the observed positive skew and the inconspicuous results for $p = 1$. Nevertheless, a double-digit pass rate and only moderately elevated excess kurtosis ($< 2$ in the large majority of cases) indicate that departures from normality are not severe.

We conclude that the claim of normality of $Z_{\mathrm{pos}}$ is reasonably well supported. This is important for the analysis of heterogeneity between covers in $Z_{\mathrm{cover}}$, as the Gaussian exponential tails imply that there is rapid convergence of the empirical mean to the true mean. The average of $\iota = 200$ data points should reduce the within-image error by a factor of $\sqrt{200} \approx 14$, which will make it small enough to ignore. In this way, the three error components $Z_{\mathrm{flips}}$, $Z_{\mathrm{pos}}$ and $Z_{\mathrm{cover}}$ can be separated.

Regarding the shape of the error distribution $Z_{\mathrm{cover}}$ for $p > 0$, we find that it is not Gaussian, but rather fat-tailed to varying extent. (We omit results of the Shapiro–Wilk test because the Gaussian null hypothesis is rejected in almost every single case.) Figure 5.4 illustrates that the error distribution found at $p = 0$ is maintained also for nontrivial embedding rates $p > 0$ and, depending on the detector, changes its shape moderately with varying embedding rate. We find good symmetry in the error distribution of RS and SPA, which allows us to omit the right tail and display more values of $p$ for

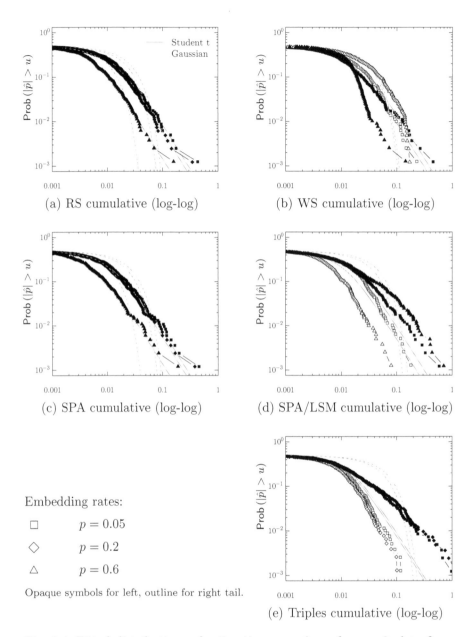

(a) RS cumulative (log-log)

(b) WS cumulative (log-log)

(c) SPA cumulative (log-log)

(d) SPA/LSM cumulative (log-log)

Embedding rates:

| | |
|---|---|
| □ | $p = 0.05$ |
| ◇ | $p = 0.2$ |
| △ | $p = 0.6$ |

Opaque symbols for left, outline for right tail.

(e) Triples cumulative (log-log)

Fig. 5.4: Fitted distributions of estimation error $\hat{p} - p$ for $p > 0$; data from 800 never-compressed greyscale images (set B); location and scale parameters estimated; $\nu = 2$

the left tail (Fig. 5.4 (a) and (c)). For $p > 0$, the shape of the left tail is more important, since far underestimation leads to false negatives or misses. The remaining three detectors exhibit more asymmetric distributions with a fatter left tail for SPA/LSM and Triples and a lighter one for WS, though only at higher embedding rates $p = 0.6$. In any case, the fat-tailed behaviour remains a matter of fact for almost all settings of $p$, and the Student $t$ distribution with $\nu = 2$ appears to be a reasonable symmetric approximation to the unknown true distribution of between-image errors $Z_{\text{cover}}$. Consequently, further analyses of the effect of heterogeneity between covers on detection performance require adequate handling of heavy tails.

Note that in practice, the change in shape and scale of $Z_{\text{cover}}$ as $p$ varies is sometimes ignored to simplify the analysis. This leads to the so-called *shift hypothesis* in [124], which postulates that the (heavy-tailed) error between images for $p = 0$ is shifted linearly by the embedding rate. This corresponds to the absolute bias model we studied in [22] along with alternative models; and it is demonstrated there that the simplification of going along with the shift hypothesis is often acceptable (and even more so if $p$ remains small).

### 5.1.3.2 Relative Size of Cover-Specific and Message-Specific Errors

The different tail characteristics of $Z_{\text{cover}}$ and $Z_{\text{message}}$ impede a direct comparison of the magnitudes of individual error components: the usual measures of dispersion (e.g., standard deviation) are not suitable for heavy-tailed distributions, and it is hard to find reasonable like-for-like comparisons between heavy-tailed and light-tailed distributions. We settled on use of a nonparametric approach and report IQRs for all distributions in Table 5.4. IQRs provide reasonable information about the spread around the centre of the distribution while not taking into account the distance of fat tails. The measures can be determined empirically for $Z_{\text{cover}}$ (after attenuating $Z_{\text{pos}}$ by averaging over $\iota = 200$ iterations) and $Z_{\text{pos}}$ (we report the median of all $N = 800$ IQRs). IQRs for $Z_{\text{flips}}$ are calculated from the quantile function, using the exact (theoretical) values for $n$ and $\frac{1}{2}p$ in a binomial CDF. Since it is known from [118] that the size of the errors depends on the image size $n$, and individual error components may exhibit varying sensitivity to $n$, we also report results from downsampled images for selected embedding rates. Note that the overall error is not equal to the sum of the individual dispersion measures as IQRs are not additive in general. The figures are merely useful to get an idea of the error magnitude and thus the relative importance of each component.[3]

---

[3] Deviations from the values plotted in Fig. 4 of [22] can be explained by the following differences in the methodology: 1) the conference paper showed results for the red channel of down-scaled colour images rather than the luminance component, 2) a different downsampling algorithm was used in [22] which resulted in smoother (and better-to-steganalyse)

Table 5.4: Error decomposition: relative size of estimation error components measured by interquartile ranges

| | Detection method | | | | | | | | | | |
|---|---|---|---|---|---|---|---|---|---|---|---|
| | RS | | WS | | SPA | | SPA/LSM | | Triples | | |
| $p$ | $Z_{cover}$ | $Z_{pos}$ | $Z_{cover}$ | $Z_{pos}$ | $Z_{cover}$ | $Z_{pos}$ | $Z_{cover}$ | $Z_{pos}$ | $Z_{cover}$ | $Z_{pos}$ | $Z_{flips}$ |
| **Full-size images** $(640 \times 457)$ | | | | | | | | | | | |
| 0.00 | 2.31 | | 2.03 | | 2.28 | | 2.34 | | 2.32 | | |
| 0.01 | 2.31 | 0.23 | 2.03 | 0.17 | 2.25 | 0.19 | 2.34 | 0.18 | 2.31 | 0.25 | 0.02 |
| 0.05 | 2.19 | 0.50 | 2.18 | 0.39 | 2.15 | 0.43 | 2.22 | 0.42 | 2.21 | 0.56 | 0.04 |
| 0.10 | 2.05 | 0.69 | 2.38 | 0.54 | 2.04 | 0.59 | 2.08 | 0.60 | 2.13 | 0.80 | 0.05 |
| 0.20 | 1.83 | 0.97 | 2.70 | 0.73 | 1.84 | 0.83 | 1.90 | 0.84 | 2.00 | 1.18 | 0.07 |
| 0.40 | 1.37 | 1.27 | 3.26 | 0.98 | 1.35 | 1.19 | 1.57 | 1.29 | | | 0.10 |
| 0.60 | 0.91 | 1.56 | 3.26 | 1.12 | 0.89 | 1.55 | 1.72 | 1.81 | | | 0.11 |
| 0.80 | 0.47 | 2.15 | 2.17 | 1.22 | 0.47 | 2.57 | | | | | 0.12 |
| 1.00 | 0.56 | 2.71 | 0.09 | 1.24 | 0.81 | 3.24 | | | | | 0.12 |
| **Downsampled to 50%** $(453 \times 324$, nearest neighbour interpolation$)$ | | | | | | | | | | | |
| 0.00 | 4.38 | | 3.28 | | 3.92 | | 3.44 | | 3.93 | | |
| 0.01 | 4.33 | 0.39 | 3.29 | 0.28 | 3.91 | 0.32 | 3.43 | 0.30 | 3.92 | 0.38 | 0.02 |
| 0.10 | 3.85 | 1.19 | 3.11 | 0.88 | 3.59 | 1.00 | 3.24 | 0.95 | 3.79 | 1.26 | 0.08 |
| 0.40 | 2.50 | 2.20 | 3.34 | 1.61 | 2.40 | 1.94 | 2.78 | 2.04 | | | 0.14 |
| 0.80 | 0.89 | 3.51 | 1.82 | 2.00 | 0.83 | 4.00 | | | | | 0.17 |
| **Downsampled to 25%** $(320 \times 229$, nearest neighbour interpolation$)$ | | | | | | | | | | | |
| 0.00 | 8.74 | | 5.32 | | 7.03 | | 5.54 | | 5.92 | | |
| 0.01 | 8.65 | 0.61 | 5.23 | 0.45 | 6.96 | 0.50 | 5.49 | 0.44 | 5.89 | 0.57 | 0.03 |
| 0.10 | 7.70 | 1.89 | 5.22 | 1.43 | 6.33 | 1.61 | 5.31 | 1.42 | 6.50 | 1.85 | 0.11 |
| 0.40 | 4.85 | 3.43 | 4.60 | 2.67 | 4.27 | 3.08 | 5.43 | 2.99 | | | 0.20 |
| 0.80 | 1.53 | 5.18 | 2.36 | 3.35 | 1.45 | 5.54 | | | | | 0.24 |

$N = 800$ never-compressed greyscale images (set B) iterated over $\iota = 200$ messages; unit of IQRs scaled to percentage points of the embedding rate.

As to the interpretation of the results, it is most obvious that the influence of $Z_{flips}$ is almost always smaller by one to two orders of magnitude than at least one other error component. We therefore conclude that, for images not much smaller than the ones in our sample, the dispersion of $Z_{flips}$ is negligible in general.[4] Moreover, between-image errors are relatively larger

images, 3) RS, SPA and Triples were harmonised to consistently use overlapping horizontal groups of pixels whereas different settings were used in [22], and 4) $Z_{message}$ has not been decomposed in the earlier publication.

[4] A sole exception worth mentioning is WS at $p = 1$, where the IQR of $Z_{flips}$ is around the same size as the IQR of $Z_{cover}$. This is, however, due to the sudden drop in the

than within-image errors for small $p$ (where estimation accuracy is crucial for low error rates in a binary stego detection scenario). This is remarkable since the problematic far tails are not even taken into account. So, in practice the relative importance of cover-specific estimation errors over message-specific ones is most likely higher than that reflected in these figures.

The size of the between-image error decreases gradually as $p$ grows for all detectors but WS, which exhibits a characteristic inverse-U relation between IQRs of $Z_{\text{cover}}$ (and other measures of dispersion [15]) and $p$. Strikingly, this pattern changes notably for the downsampled images. Later, in Chapter 6, we will elaborate further on the sensitivity of WS estimates with regard to resampling methods.

Not very surprisingly, the relative size of the large within-image error component $Z_{\text{pos}}$ grows with $p$, though it depends on the detector whether its IQRs exceed (RS, WS, SPA) or just match (SPA/LSM) the IQRs of $Z_{\text{cover}}$ for medium and large $p$. Observe also the slight tendency that the relative size of between-image errors decreases with larger images. The methods based on LSM solutions (SPA/LSM and Triples) tend to produce serious outliers for medium (Triples) and high (SPA/LSM) embedding rates. This explains the blank cells for these detectors. Any practical application of these methods requires a pre-screening with a more robust method (RS, WS, SPA) to obtain a more precise estimate in those occasions where SPA/LSM or Triples performs better than the robust methods (e.g., small or preprocessed images, not visible in the table). We refrain from interpreting the results to compare the performance between detectors as some of them do not operate with optimal settings. For example, SPA was intentionally set to use only horizontal pairs in this series of experiments to allow best comparability with RS, and further improvements of WS (to be presented in Chapter 6) had not been discovered at the time we conducted this research on cover-specific determinants.

Summing up, we can conclude that in the large majority of cases, the cover-specific estimation error is the most important source of error. This can be framed as motivation for an instance of the mixture cover model, in which heterogeneity between covers materialises in the realisation of the between-image error. Steganalytic reliability can thus benefit most from ways to predict size and, if possible, direction of this error component from cover properties. Any such approach should take into account the fat-tailed characteristic of this component.

## 5.2 Measurement of Sensitivity to Cover Properties

The random estimation error $Z_{\text{cover}}$ can be interpreted as an unobservable random variable in the mixture cover model of Sect. 3.3.1. As argued there,

---

between-image error of WS when $p$ approaches 1. Nevertheless, $Z_{\text{pos}}$ remains ten times larger than $Z_{\text{flips}}$ even in this case.

the reliability of steganalysis can be improved if (part of) the variation of $Z_{cover}$ can be predicted for a realisation in a concrete instance, for example, from macroscopic cover properties which are approximately invariant to the embedding function (cf. Eq. (3.25)). In the next section we recall regression analysis as the method of choice to study multivariate causal relations between random variables and discuss requirements in the specific case of quantitative steganalysis with heavy-tailed error distribution. Sections 5.2.2 and 5.2.3 report example results for factors influencing detection performance in terms of between-image and within-image errors, respectively.

## 5.2.1 Method

Regression analysis is a standard statistical tool to estimate the relation between a *response* variable of interest and one or more predictors as *explanatory variables*.[5] In its simplest linear form,

$$y_i = b_0 + b_1 v_{i,1} + \cdots + b_k v_{i,k} + \sigma \epsilon_i, \tag{5.9}$$

the response $y$ is approximated by $k$ explanatory variables in the columns of matrix $v$ [6] via the vector of linear *regression coefficients* $b$ plus a not linearly explainable residual $\epsilon$. In matrix notation we can write

$$y = v b + \sigma \epsilon, \tag{5.10}$$

where, by convention, matrix $v$ has $k+1$ columns and all elements of the leftmost column, indexed by 0, equal to 1. The $k+1$ regression coefficients $\hat{b}$ can be estimated from data ($1 \le i \le N$). If $\epsilon$ is IID Gaussian, $\epsilon_i \sim \mathcal{N}(0,1)$, then, according to the Gauss-Markov theorem, the maximum likelihood estimate concurs with the least-squares solution,

$$\hat{b} = \arg\min_b \sum_{i=1}^{n} \epsilon_i^2 = \arg\min_b \sum_{i=1}^{n} \left( y_i - \sum_{j=0}^{k} b_j v_{i,j} \right)^2. \tag{5.11}$$

The normal equations of this linear least-squares problem in matrix notation,

$$\left( v^\mathsf{T} v \right) \hat{b} = v^\mathsf{T} y, \tag{5.12}$$

---

[5] Alternative terms for the response in the literature include 'dependent variable' or 'endogenous variable'. Similarly, explanatory variables are also referred to as 'independent variables', 'exogenous variables' or simply 'predictors'. We shall use the latter for brevity where space matters.

[6] We depart from the use of the more common notation $x$ for the matrix of explanatory variables as this symbol is reserved for cover and stego objects.

can be solved for $\hat{\boldsymbol{b}}$ if $\boldsymbol{v}^{\mathsf{T}}\boldsymbol{v}$ is not singular, as follows:

$$\hat{\boldsymbol{b}} = \left(\boldsymbol{v}^{\mathsf{T}}\boldsymbol{v}\right)^{-1}\boldsymbol{v}^{\mathsf{T}}\boldsymbol{y}. \tag{5.13}$$

This opens the toolbox of statistical inference with linear models: the sign and size of coefficients $b_j$ can be interpreted as measures for direction and strength of dependence between the explanatory variable $\boldsymbol{v}_j$ and response $\boldsymbol{y}$. Hypothesis tests based on the asymptotic of estimates $\hat{b}_j$ can tell statistically significant coefficients (i.e., $\mathsf{Prob}(\hat{b}_j \mid b_j = 0)$ below a generally accepted significance level, e.g., 0.1%) from spurious nonzero coefficients.

A direct application of this standard method to explain estimation errors $y_i = \hat{p}_i - p$ as linear functions of cover properties $\mathsf{Proj}(\boldsymbol{x}_i^{(0)})$ is impeded by the violation of the Gaussian IID assumption for $\epsilon_i$. Instead, we have to resort to heteroscedastic Student $t$ regression, which allows for two separate linear models for location and scale parameters:

$$y_i = a_0 + a_1 u_{i,1} + \cdots + a_k u_{i,k_u} + \sigma_i \epsilon_i, \qquad \epsilon_i \sim t(0,1,\nu) \tag{5.14}$$
$$\log(\sigma_i^2) = b_0 + b_1 v_{i,1} + \cdots + b_k v_{i,k_v}. \tag{5.15}$$

The transformation of the scale $\sigma$ in Eq. (5.15) ensures that it remains always positive.

The gain in flexibility and looser assumptions come at the cost of a more complex (and, for small $N$, less stable) estimation procedure. The scoring method by Taylor and Verbyla [224] to solve the maximum likelihood problem in Eqs. (5.4) and (5.5) accepts explanatory variables for both location ($\boldsymbol{u}$) and scale ($\boldsymbol{v}$) parameters and shall be used to measure the effect of two exemplary image properties, local variance and saturation, on the error distribution. The macroscopic properties are defined as follows.

Local variance (ignoring boundary conditions to simplify the notation):

$$\mathsf{LocVar}(\boldsymbol{x}) = \frac{1}{2n}\sum_{i=1}^{n}\left[\left(x_i - x_{\mathsf{Left}(i)}\right)^2 + \left(x_i - x_{\mathsf{Upper}(i)}\right)^2\right] \tag{5.16}$$

Saturation (for eight-bit signals):

$$\mathsf{Sat}(\boldsymbol{x}) = \frac{1}{n}\sum_{i=1}^{n}\left(\delta_{x_i,0} + \delta_{x_i,255}\right) \tag{5.17}$$

A visual inspection of the distribution of the property measurements between images showed a lognormal-like distribution. So we transformed the measures by taking the natural logarithm[7] before entering them as explanatory variables in our regression analysis. All models were fitted to $N = 800$ data points from *different* images to ensure independence of the residuals. Image set B (cf. Appendix A) serves as reference for all experiments in this chapter, mainly because image set A contains too few saturated images to estimate the influence of this property (see Table A.1).

## 5.2.2 Modelling the Shape of the Between-Image Distribution

To study the individual and combined influence of image properties on the estimation error, seven specifications of a heteroscedastic $t$ regression model have been fitted for each of five quantitative detectors.[8] Specifications $\langle 1 \rangle$–$\langle 4 \rangle$ are fitted to the empirical distribution of $Z_{\text{cover}}$, after attenuating the effect of $Z_{\text{message}}$ by averaging, for $p = 0$. In specifications $\langle 5 \rangle$–$\langle 7 \rangle$ we control for the influence of the embedding rate and fit to the empirical distribution of $Z_{\text{cover}}$, in which objects with $p \in \{0, 0.01, 0.05, 0.1, 0.2, 0.4, 0.6, 0.8\}$ occur with equal probability. Due to the performance drop of the LSM methods for high embedding rates, a reduced set of embedding rates has been used for these methods: up to (and including) $p = 0.6$ for SPA/LSM and $p = 0.2$ for Triples, respectively. The rationales behind the choice of specifications are as follows:

$\langle 1 \rangle$ This specification is the default model without any predictors. Only the constant terms of the location ($a_0$) and scale ($b_0$) model are estimated. The coefficient values correspond to the parametric performance measures reported in Table 5.2 (note the different scaling for the coefficients of the scale model).

$\langle 2 \rangle$ This specification models the bivariate influence of local variance on $Z_{\text{cover}}$. Corresponding scatter plots with fitted regression lines superimposed are depicted in the left column of Fig. 5.5.

$\langle 3 \rangle$ This specification models the bivariate influence of saturation on $Z_{\text{cover}}$. The corresponding scatter plots and regression lines are shown in the right column of Fig. 5.5.

$\langle 4 \rangle$ This specification models the joint influence of local variance and saturation. By comparing the coefficient values to $\langle 2 \rangle$ and $\langle 3 \rangle$, one can judge if the influence from both properties is roughly additive. We are not in a position to fit a model with an interaction term (the proper way to test for additivity) because saturation is constant at zero for more than 50% of the images in the sample. This leads to a difficult nonlinear relation with

---

[7] A small offset of $10^{-5}$ has been added to the saturation measures to avoid undefined results for images without any saturated pixel.

[8] We use angle brackets '$\langle \rangle$' to enumerate the specifications.

Table 5.5: Regression coefficients fitted to $Z_{\text{cover}}$ of RS analysis

| Predictor | | | | Specification | | | |
|---|---|---|---|---|---|---|---|
| | $\langle 1 \rangle$ | $\langle 2 \rangle$ | $\langle 3 \rangle$ | $\langle 4 \rangle$ | $\langle 5 \rangle$ | $\langle 6 \rangle$ | $\langle 7 \rangle$ |
| **Location model** | | | | | | | |
| *constant* | 0.18** (0.063) | −0.58 (0.332) | 1.54*** (0.315) | 0.98* (0.475) | 0.68*** (0.078) | 0.63*** (0.177) | 0.39 (0.461) |
| local variance (log) | | 0.14* (0.064) | | 0.08 (0.063) | | 0.01 (0.031) | 0.05 (0.087) |
| saturation (log) | | | 0.13*** (0.030) | 0.12*** (0.029) | | | |
| emb. rate $p$ | | | | | −0.03 (0.121) | −0.01 (0.117) | 0.35 (0.712) |
| loc. var. (log) $\times\ p$ | | | | | | | −0.07 (0.135) |
| **Scale model** | | | | | | | |
| *constant* [a] | −8.55 (0.079) | −11.43 (0.584) | −7.38 (0.307) | −10.05 (0.685) | −8.35 (0.109) | −10.88 (0.593) | −10.90 (0.877) |
| local variance (log) | | 0.52*** (0.106) | | 0.46*** (0.106) | | 0.46*** (0.106) | 0.46** (0.158) |
| saturation (log) | | | 0.13*** (0.030) | 0.11*** (0.031) | | | |
| emb. rate $p$ | | | | | −3.65*** (0.281) | −3.60*** (0.282) | −3.62 (2.034) |
| loc. var. (log) $\times\ p$ | | | | | | | 0.00 (0.369) |

$N = 800$ never-compressed grey scale images (set B); std. errors in brackets; coefficients of location model scaled to percentage points of embedding rate; $\nu = 2$; significance levels: *** $\leq 0.001$, ** $\leq 0.01$, * $\leq 0.05$.
[a] no significance test computed due to lack of null hypothesis

partial multicollinearity between the interaction term and the individual properties.

$\langle 5 \rangle$ This specification models the bivariate influence of the net embedding rate $p$ on $Z_{\text{cover}}$. Since the analysis is bivariate, scatter plots have been prepared and can be seen in Fig. 5.6 (a)–(e).

$\langle 6 \rangle$ This specification models the joint influence of local variance and embedding rate on $Z_{\text{cover}}$. By comparing the coefficient values to $\langle 2 \rangle$, one can judge if the direction and strength of influence for covers ($p = 0$) broadly remains the same for stego objects (invariance of influence to embedding). A similar specification for saturation and embedding rate has been considered, but abandoned due to estimation difficulties. Saturation is per se negatively related to the embedding rate, which is a source of multicollinearity.

$\langle 7 \rangle$ The last specification extends $\langle 6 \rangle$ by an interaction term. It allows us to measure the influence of local variance conditionally on the embedding rate (i.e., does the disturbing effect of local variance around $p = 0$ fade or intensify when $p$ grows?). Both explanatory variables are reasonably independent and balanced over the sample so that the model with an interaction term could be fitted without statistical difficulties.

Estimated coefficients for all specifications and RS analysis as detector are reported in Table 5.5. Similar tables for the remaining four detectors have been placed in Appendix G, Tables G.1, G.3, G.5 and G.7. Coefficients annotated with asterisks differ statistically significantly from 0 (using the test statistic and standard error estimates of [224]).

Significant coefficients in the constant of the *location model* indicate a bias on the detector. If further significant coefficients show up in the other predictors of the location model, then the sign decides whether the bias strengthens (same sign) or is compensated (different sign) as the predictor increases. For example, RS has a positive bias (i.e., it overestimates around $p = 0$), and tends to overestimate more the higher the proportion of saturated pixels in a cover. An ideal detector should be bias-free.

Positive significant coefficients in the predictors of the *scale model* indicate that the dispersion of $Z_{\mathrm{cover}}$ increases, i.e., the detector accuracy deteriorates, as the value of the explanatory variable grows. Conversely, negative significant coefficients signal a performance increase for higher values of the explanatory variable. For example, RS analysis, on average, loses precision both with growing local variance and increasing saturation. This holds independently (see $\langle 2 \rangle$, $\langle 3 \rangle$ and Fig. 5.5 (a)–(b)), as well as jointly ( $\langle 4 \rangle$).

The results from 70 regression equations are too numerous, and partly not sufficiently relevant, to discuss them in all detail. We refer the reader to the scatter plots to get an impression of the type of evidence this method provides, and comment only the most remarkable findings. In addition, Table 5.7 summarises qualitative information on strength and direction of influencing factors for all tested detection methods.

In general, the results indicate measurable influence from the tested predictors, but their partial contributions to detection accuracy are quite independent. Neither does the size of the coefficients change substantially between estimates for specifications $\langle 2 \rangle$ and $\langle 4 \rangle$ (local variance) or $\langle 3 \rangle$ and $\langle 4 \rangle$ (saturation), nor can we observe significant interaction terms in specification $\langle 7 \rangle$, except for WS (cf. Tab. G.1 in Appendix G). WS, however, is a special case in this respect. The linear models do not appropriately capture its typical inverse-U shape (cf. Sect. 5.1.3.2 above) of detection accuracy as a function of $p$, which explains the difficult-to-interpret coefficients of specifications $\langle 5 \rangle$–$\langle 7 \rangle$. We have also fitted models with quadratic terms of the embedding rate as an additional predictor, and Fig. 5.6 (f) illustrates that the characteristic shape is well recovered in both the scale term and the location term. We refrain from reporting detailed coefficients because they are not comparable to other detectors. Overall, Triples appears least affected by saturation and the embedding rate. While the case for saturation is true, the results for the embedding rate need more careful interpretation. Note that both Triples and SPA/LSM fail for higher embedding rates and therefore the regression models

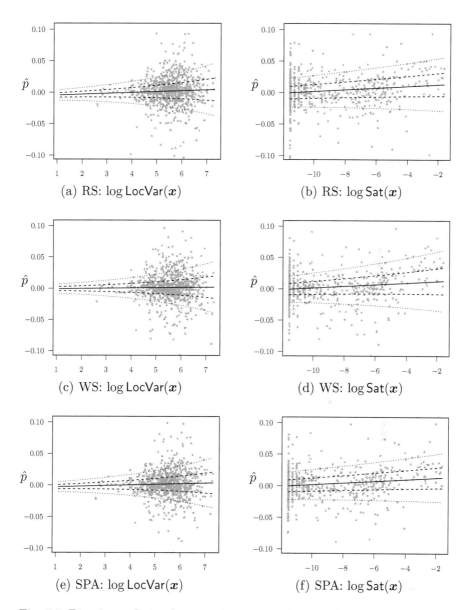

Fig. 5.5: Bivariate relation between image properties and estimation error at $p = 0$ for $N = 800$ never-compressed greyscale images (set B); heteroscedastic $t$ regression lines (solid for location $\mu$, dashed for quartiles, dotted for 10–90% quantile); $\nu = 2$

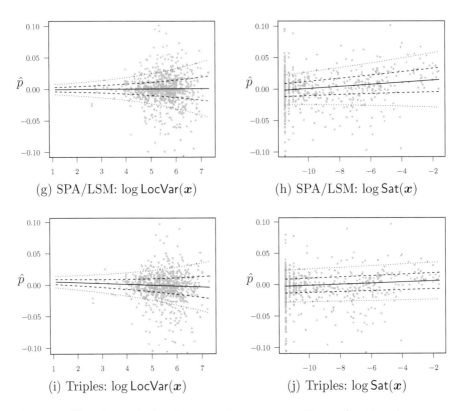

(g) SPA/LSM: $\log \mathsf{LocVar}(\boldsymbol{x})$    (h) SPA/LSM: $\log \mathsf{Sat}(\boldsymbol{x})$

(i) Triples: $\log \mathsf{LocVar}(\boldsymbol{x})$    (j) Triples: $\log \mathsf{Sat}(\boldsymbol{x})$

Fig. 5.5: Bivariate relation between image properties and estimation error (continued); see caption on p. 147 for details

were fitted on a reduced range. So the results of specifications $\langle 5 \rangle$–$\langle 7 \rangle$ are not directly comparable to RS, WS and SPA (between which a comparison is possible).

## 5.2.3 Modelling the Shape of the Within-Image Distribution

In Sect. 5.1.3.2 we found that $Z_{\mathrm{cover}}$ is the quantitatively most important source of detection error. Further, theoretical arguments suggest that $Z_{\mathrm{flips}}$ should be independent of image properties, so $Z_{\mathrm{pos}}$ remains to be studied.

Although the realisation of the within-image error component $Z_{\mathrm{pos}}$ is a function of the message, its relation to image properties is not entirely beyond

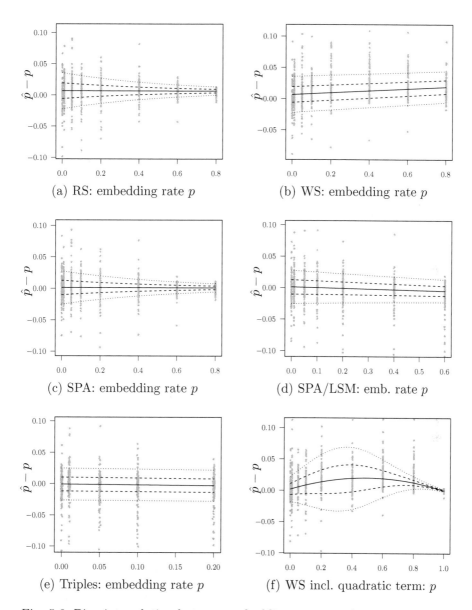

Fig. 5.6: Bivariate relation between embedding rate $p$ and estimation error $\hat{p} - p$; $N = 800$ never-compressed greyscale images (set B); heteroscedastic $t$ regression lines (solid for location $\mu$, dashed for quartiles, dotted for 10–90% quantile); $\nu = 2$

Table 5.6: Regression coefficients fitted to $\log \hat{\sigma}_i$ of $Z_{\text{pos}}$ for RS analysis

| Predictor | Specification | | | | | | |
|---|---|---|---|---|---|---|---|
| | $\langle 1 \rangle$ | $\langle 2 \rangle$ | $\langle 3 \rangle$ | $\langle 4 \rangle$ | $\langle 5 \rangle$ | $\langle 6 \rangle$ | $\langle 7 \rangle$ |
| *constant* [a] | $-5.56$ | $-7.86$ | $-5.77$ | $-8.27$ | $-5.65$ | $-7.50$ | $-7.89$ |
| | (0.018) | (0.103) | (0.069) | (0.118) | (0.028) | (0.124) | (0.199) |
| local variance (log) | | 0.42 *** | | 0.43 *** | | 0.34 *** | 0.41 *** |
| | | (0.019) | | (0.018) | | (0.022) | (0.036) |
| saturation (log) | | | $-0.02$ ** | $-0.03$ *** | | | |
| | | | (0.007) | (0.005) | | | |
| emb. rate $p$ | | | | | 1.94 *** | 1.97 *** | 2.83 *** |
| | | | | | (0.054) | (0.047) | (0.343) |
| loc. var. (log) $\times$ $p$ | | | | | | | $-0.16$ * |
| | | | | | | | (0.062) |
| **R-squared (adj.)** | | 0.39 | 0.01 | 0.42 | 0.62 | 0.70 | 0.71 |

$N = 800$ never-compressed grey scale images (set B); std. errors in brackets; specifications $\langle 1 \rangle$–$\langle 4 \rangle$ fitted for $p = 0.05$; significance levels: *** $\leq 0.001$, ** $\leq 0.01$, * $\leq 0.05$.
[a] no significance test computed due to lack of null hypothesis

our focus on heterogeneity *between covers*. This is so because the dispersion of the within-image error component $Z_{\text{pos}}$ in fact varies between images. So we fit single regression equations that explain the empirical standard deviation $\hat{\sigma}$ of $Z_{\text{pos}}$ with the same set of explanatory variables regressed on $Z_{\text{cover}}$. It has been shown above that $Z_{\text{pos}}$ is reasonably close to a Gaussian distribution to allow inference via second moment statistics. Incidentally, the distribution of $\hat{\sigma}_i$ over $i = 1, \ldots, N$ images in set B turned out to fit a lognormal distribution sufficiently well.[9] So we decided to take the logarithm of the response and fit simple homoscedastic ordinary least-squares (OLS) regression of the form of Eq. (5.9) for all seven specifications described in the previous section. To be on the safe side in case of potentially influential outliers, we reproduced our results using a robust M-estimate fitted with iterated reweighted least squares (IWLS) [106, 229]. As both methods come to virtually the same results, we report only the OLS estimates. Figure G.2 in Appendix G summarises all bivariate scatter plots with both OLS and robust regression lines superimposed.

In general, the results do not differ substantially between detectors, so we report the coefficients for RS analysis in Table 5.6 and comment on the other methods only informally. The corresponding tables for the other methods can be found in Appendix G (Tables G.2, G.4, G.6 and G.8). Note that we report so-called *R-squared* measures of model fit, adjusted for the number of estimated coefficients across specifications. This metric reports the fraction

---

[9] This finding does not necessarily generalise to other cover sources. We have not validated it on other image sets than set B because calculating parameters of the within-image error is computationally very expensive (in the order of weeks of processor time) and the expected results are only of minor relevance.

of explained variance with respect to the default specification $\langle 1 \rangle$.[10] Independently of the detection methods, the amount of local variance is positively associated with the scale of the within-image error component $Z_{\text{pos}}$. Conversely, a higher proportion of saturated pixels, on average, goes along with smaller within-image errors. However, saturation as a predictor is much less influential (though still statistically significant for all methods but WS) than local variance. Finally, as theoretically justified, the embedding rate is positively associated with the dispersion of $Z_{\text{pos}}$. The estimated coefficients also reveal that in fact the embedding rate has more explanatory power than both image properties (compare the $R$-squared values between specifications). The negative and significant interaction terms for RS and SPA indicate that the influence of local variance as a predictor slightly fades with increasing embedding rate. This property is not shared by the other detectors.

## 5.3 Summary and Conclusion

In this chapter, we have developed statistical models to explore heterogeneity between covers for the class of quantitative detectors of LSB replacement steganography. When fitted to data from stego detection simulations, these models allow us to derive conclusive performance metrics and offer a methodology to identify and measure the influence of cover-specific properties on detection performance. Two exemplary image properties, local variance and saturation, have been studied for five common detection methods in conjunction with the embedding rate as the third explanatory variable.

Several results convey implications for broader contexts of steganalysis research:

1. We have identified a fat-tailed behaviour of the between-image error distribution, in particular one where the second moment (variance) is not necessarily finite. This finding puts a cautionary warning on attempts to model heterogeneity between images with standard statistical tools, which almost always require finite variance. This already affects summary statistics of steganalysis results over a test set of images. As an appropriate countermeasure, we have proposed alternative parametric and nonparametric measures which do not suffer from problems of infinite variance.
2. We have decomposed the estimation error into one image-specific and two message-specific (i.e., within-image) components and measured their relative contribution to the overall estimation error. Although the within-image error is often treated as marginal and thus disregarded, we have found that there exist situations in which it should not be ignored. In

---

[10] Although a number of pseudo-$R$-squared metrics for other regression models than Gaussian OLS can be found in the literature, we are not aware of a similarly telling and reliable summary metric for the case of heteroscedastic Student $t$ regression.

particular, when modelling heterogeneity between images, the within-image errors appear as convoluted noise and blur the heterogeneity. As a consequence, we have proposed a method to eliminate or attenuate the distortion due to within-image error components. A drawback of the proposed method is that it involves computationally very expensive simulations.

3. The distribution of the within-image error itself varies between images and should be regarded when studying heterogeneity.

4. We have applied heteroscedastic Student $t$ regression models (for the between-image error) and homoscedastic ordinary least-squares regression (for the scale of the between-image error) to identify causal relations between macroscopic image properties and detection performance. The gist of the results from altogether 105 regression equations is summarised qualitatively in Table 5.7.

Table 5.7: Summary of factors influencing detection accuracy

| | Between-image error | Between-image error | Within-image error |
|---|---|---|---|
| Influencing factor | bias | dispersion | dispersion |
| **local variance** | positive relation for RS and negative relation for Triples | positive relation for all detectors | positive relation for all detectors |
| **saturation** | positive relation for all detectors | positive relation for all detectors but Triples | small negative relation for all detectors but WS |
| **emb. rate** | negative relation for SPA/LSM, nonlinear (inv. U-shape) relation for WS | negative relation for RS, SPA(/LSM), nonlinear relation (inv. U-shape) for WS, little influence on Triples | positive relation for all detectors |

It was not the objective of this chapter to come up with direct recommendations for improved detectors based on the findings presented. To come to sound conclusions, such endeavours need to be targeted to individual detectors. In this case, the methods can be applied to detector-specific quantities (e.g., using parity co-occurrence as predictor for WS accuracy instead of saturation, or compliance with cover assumptions for RS and SPA). Moreover, when quantitative detectors are to be applied to generate binary steganalysis decisions, the decision threshold $\tau$ can be adapted to the dispersion of the estimation error, as predicted from macroscopic properties of the concrete suspect object under analysis. This should reduce overall error rates. Similarly, predictable bias could be corrected to improve estimation accuracy and reduce error rates in binary steganalysis decisions (discriminators).

Another application for the proposed methods is steganalysis benchmarking. The parametric models allow us to draw statistical inference in terms of hypothesis tests to decide about the statistical significance of performance differentials between (variants of) detectors. This aspect is not further investigated here due to our focus on heterogeneity between covers.

As to the limitations, the regression models proposed in this chapter are generic (or 'reduced-form' in the terminology of econometrics) specifications imposed on simulated data. This means that the models do not reflect possible theoretical links between the predictors and the error distributions. The only way to discover such links are complicated analytic derivations (see [125] for an analysis of estimation errors of a simplified version of SPA/LSM). So-obtained functional forms are usually detector-specific. By contrast, our method allows for comparisons between detectors, which is essential for benchmarking applications.

The methods and procedures described here for LSB replacement steganalysis in the spatial domain can be generalised to other embedding operations and domains—a first adaptation to JSteg detectors can be found in [237]; but care must be taken. For example, (modified) LSB replacement in JPEG images has to control the varying numbers of nonzero DCT coefficients between images as a fourth error component.

# Chapter 6
# Improved Weighted Stego Image Steganalysis

This chapter builds on the weighted stego image (WS) steganalysis method, a quantitative detector for LSB replacement steganography invented by Fridrich and Goljan [73]. Section 2.10.4 above contains a detailed description of the method in its original form (hence called 'standard WS'). A special property of WS is its explicit image model, represented by the pair of functions Pred and Conf, which are employed to estimate the cover signal. This makes the method a particularly interesting object of study from the point of view of our theory of cover models outlined in Chapter 3.

The research contribution of this chapter is divided into two main sections. Firstly, in Sect. 6.1, we propose an improved cover model for WS, demonstrate its performance gains compared to standard WS, and identify which cover sources lead to advantages of individual improvements of the cover model. This creates the link to the heterogeneity of cover sources and conditional cover models.

Secondly, Sect. 6.2 focuses on problems due to heterogeneous covers. We present a specialised WS cover model for JPEG pre-compressed (as opposed to never-compressed) covers and thereby demonstrate that the motivating example for our mixture cover model (cf. Sect. 3.3), in which preprocessing is the unobservable random variable, in fact helps to improve steganalysis in practice.

## 6.1 Enhanced WS for Never-Compressed Covers

Recall from Sect. 2.10.4 the WS detector, Eqs. (2.40) to (2.43): the main estimation equation,

$$\hat{p} = 2 \sum_{i=1}^{n} w_i \left( x_i^{(p)} - \overline{x}_i^{(p)} \right) \left( x_i^{(p)} - \mathsf{Pred}(\boldsymbol{x}^{(p)})_i \right), \qquad (6.1)$$

the weights formula with constant $u = 1$ for standard weighted WS,

$$w_i = \frac{\vartheta}{u + \mathsf{Conf}(\boldsymbol{x}^{(p)})_i}, \tag{6.2}$$

and the (reciprocal) normalisation factor for all weights,

$$\frac{1}{\vartheta} = \sum_{i=1}^{n} \frac{1}{u + \mathsf{Conf}(\boldsymbol{x}^{(p)})_i}. \tag{6.3}$$

For unweighted WS, Eq. (6.2) simply becomes $w_i = n^{-1}$. Fridrich and Goljan [73] have proposed two-dimensional linear filters of the form

$$\begin{bmatrix} 0 & \frac{1}{4} & 0 \\ \frac{1}{4} & 0 & \frac{1}{4} \\ 0 & \frac{1}{4} & 0 \end{bmatrix} \tag{6.4} \qquad \text{and} \qquad \begin{bmatrix} \frac{1}{8} & \frac{1}{8} & \frac{1}{8} \\ \frac{1}{8} & 0 & \frac{1}{8} \\ \frac{1}{8} & \frac{1}{8} & \frac{1}{8} \end{bmatrix} \tag{6.5}$$

to be used in function Pred to estimate the value of the centre pixel from its neighbours. Function Conf is defined as the empirical variance of all neighbours with nonzero filter coefficient (cf. Eq. (2.43)).

Standard weighted WS has been known as moderately good detector, which had a slight advantage over structural detectors at high net embedding rates $p > 0.7$,[1] but underperformed for relevant ranges of $p$ close to 0 [22, 73]. So the method remained widely unknown for long and attracted scientific interest only for its simplicity and mathematical beauty. However, the relative performance disadvantage can largely be attributed to a suboptimal image model, and consequent model refinements can leverage the method to a level playing field with other quantitative detectors for LSB replacement in never-compressed covers (and even beyond). We first explain the refinements, one by one, in the following subsections, before we evaluate the achievable performance improvements experimentally in Sect. 6.1.4.

### 6.1.1 Enhanced Predictor

Experimental evidence in [73] suggests that the linear filter in Eq. (6.4) yields better detection performance than the filter in Eq. (6.5). At first sight, this result is puzzling, as one would expect that incorporating *more* information in Pred should lead to more stable predictions due to decreasing standard errors, and thus, on average, to a better approximation of the cover $\boldsymbol{x}^{(0)}$, thereby *boosting* performance. However, more information does not yield better predictions if the available information is combined in a suboptimal way.

---

[1] Cf. Fig. 5.6 (f) in Chapter 5 for the performance of standard WS for very high $p$.

We experimentally found that a linear predictor using all eight neighbours performs substantially better if the linear filter is configured as follows:

$$\begin{bmatrix} -\frac{1}{4} & \frac{1}{2} & -\frac{1}{4} \\ \frac{1}{2} & 0 & \frac{1}{2} \\ -\frac{1}{4} & \frac{1}{2} & -\frac{1}{4} \end{bmatrix}. \tag{6.6}$$

Although we are not able to analytically show that this predictor is necessarily better for empirical images, the following considerations provide better intuition into the sign structure of the filter coefficients in Eq. (6.6). For this purpose, let us temporarily reduce the dimensionality of the problem from eight to three explanatory variables, on which we impose a simple constant-correlation image model.

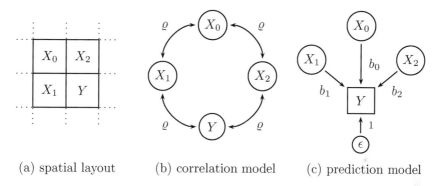

(a) spatial layout          (b) correlation model          (c) prediction model

Fig. 6.1: Image model for derivation of enhanced predictor coefficients

Consider a neighbourhood of four pixels, arranged in a $2 \times 2$ matrix as in Fig. 6.1 (a). Realisations of the random variables $(X_0, X_1, X_2, Y)$ are drawn from a zero-mean unit-variance multivariate Gaussian distribution with pairwise correlation $\varrho$ between directly adjacent pixels, that is, in vertical and horizontal directions. Spatial correlation is a reasonable assumption for typical image data, whereas the multivariate Gaussian assumption only holds in certain areas of natural images [221]. Nevertheless, we make this assumption here for simplicity. Figure 6.1 (b) visualises the correlation relations and Fig. 6.1 (c) depicts the corresponding linear regression model to predict the intensity of realisation $y$ from the realisations of its neighbours $(x_0, x_1, x_2)$. The graphs are inspired by the visualisation conventions for structural equation models [27]: nodes correspond to random variables, two-headed arrows denote correlation relations, and single-headed arrows point from explanatory (circle nodes) to response (square nodes) variables in linear regressions. Node $\epsilon$ is the error term capturing unexplained variance.

Finding the optimal linear filter coefficients in this setting is equivalent to finding the regression weights $(b_0, b_1, b_2)$, where $b_1 = b_2$ due to symmetry. The structure of Fig. 6.1 (b) implies a correlation matrix of the form

$$\boldsymbol{\Sigma} = \mathsf{E}\left(\boldsymbol{X}^\mathsf{T}\boldsymbol{X}\right) = \begin{bmatrix} 1 & \varrho & \varrho & 0 \\ \varrho & 1 & 0 & \varrho \\ \varrho & 0 & 1 & \varrho \\ 0 & \varrho & \varrho & 1 \end{bmatrix} \quad \text{with} \quad \boldsymbol{X}^\mathsf{T} = \begin{bmatrix} X_0 \\ X_1 \\ X_2 \\ Y \end{bmatrix}. \tag{6.7}$$

Correlated multivariate Gaussian random variables can be obtained by 'mixing' a row vector of independent Gaussian random variables with an upper triangle mixing matrix $\boldsymbol{a}$,

$$\boldsymbol{X} \sim \mathcal{N}(\boldsymbol{0}, \boldsymbol{\Sigma}) = \boldsymbol{R}\boldsymbol{a} \quad \text{where elements } R_{1,i} \sim \mathcal{N}(0, 1). \tag{6.8}$$

After inserting Eq. (6.8) into Eq. (6.7), we obtain

$$\boldsymbol{\Sigma} = \mathsf{E}\left(\boldsymbol{X}^\mathsf{T}\boldsymbol{X}\right) = \mathsf{E}\left((\boldsymbol{R}\boldsymbol{a})^\mathsf{T}\boldsymbol{R}\boldsymbol{a}\right) = \boldsymbol{a}^\mathsf{T}\underbrace{\mathsf{E}\left(\boldsymbol{R}^\mathsf{T}\boldsymbol{R}\right)}_{=I}\boldsymbol{a} = \boldsymbol{a}^\mathsf{T}\boldsymbol{a}. \tag{6.9}$$

Hence, for every positive definite correlation matrix $\boldsymbol{\Sigma}$, the mixing matrix $\boldsymbol{a}$ can be found by the Cholesky decomposition, for which efficient numerical and symbolic algorithms exist (e.g., [223]). To rewrite the correlation model in Fig. 6.1 (b) as prediction model (c), it is convenient so split up matrix $\boldsymbol{a}$ as follows:

$$\boldsymbol{a} = \begin{bmatrix} a_{1,1} & a_{1,2} & a_{1,3} & a_{1,4} \\ 0 & a_{2,2} & a_{2,3} & a_{2,4} \\ 0 & 0 & a_{3,3} & a_{3,4} \\ 0 & 0 & 0 & a_{4,4} \end{bmatrix} = \begin{bmatrix} \begin{bmatrix} \boldsymbol{a}_\diamond \end{bmatrix} & \begin{bmatrix} \boldsymbol{v} \end{bmatrix} \\ \begin{bmatrix} 0 & 0 & 0 \end{bmatrix} & a_{4,4} \end{bmatrix}. \tag{6.10}$$

Since $a_{4,i} = 0 \; \forall i \neq 4$, Eq. (6.8) can be reduced to the top rows that explain $X_i$ of $\boldsymbol{X}$,

$$\begin{bmatrix} X_0 & X_1 & X_2 \end{bmatrix} = \begin{bmatrix} R_0 & R_1 & R_2 \end{bmatrix} \boldsymbol{a}_\diamond, \tag{6.11}$$

and then solved to find the realisations $(r_0, r_1, r_2)$ for observed neighbour pixels $(x_0, x_1, x_2)$:

$$\boldsymbol{r} = \begin{bmatrix} r_0 & r_1 & r_2 \end{bmatrix} = \begin{bmatrix} x_0 & x_1 & x_2 \end{bmatrix} \boldsymbol{a}_\diamond^{-1} = \boldsymbol{x}\,\boldsymbol{a}_\diamond^{-1}. \tag{6.12}$$

From Eqs. (6.8) and (6.11), we obtain a linear predictor for $Y$,

$$Y = \boldsymbol{r}\boldsymbol{v} + a_{4,4} \cdot R_3 = \boldsymbol{x}\,\boldsymbol{a}_\diamond^{-1}\boldsymbol{v} + \epsilon = \boldsymbol{x}\boldsymbol{b} + \epsilon, \tag{6.13}$$

so the linear filter coefficients are given by

$$\begin{bmatrix} b_0 \\ b_1 \\ b_2 \end{bmatrix} = a_\diamond^{-1} v. \tag{6.14}$$

The symbolic solution of our $2 \times 2$ model, as detailed in Appendix D, is

$$b_0 = -\frac{2\varrho^2}{1 - 2\varrho^2} \tag{6.15}$$

$$b_1 = b_2 = \frac{\varrho}{1 - 2\varrho^2}. \tag{6.16}$$

Observe that $b_0$ is negative for all permissible $\varrho < \frac{1}{2}$, which suggests that corner coefficients of a linear prediction filter should be negative to offset the indirect relation of diagonal pixels via both horizontal and vertical neighbours. Cases where $\varrho \geq \frac{1}{2}$ do not exist, as the resulting correlation matrix would not be positive definite.

So far, we have worked in a zero-mean model, where the term $\epsilon = a_{4,4} \cdot R_4$ can be ignored because $\mathsf{E}(\epsilon) \propto \mathsf{E}(R_4) = 0$. If we relax this assumption, $b$ has to be scaled to $\sum_i b_i = 1$ to ensure that the linear predictor preserves the mean (all practical WS predictors do so). Note that such scaling does not alter the signs of the coefficients.

The procedure described so far can be equally applied to larger neighbour-hoods, although the symbolic Cholesky decomposition becomes increasingly complicated. Figure 6.2 shows the numerical solutions for Eq. (6.14) and subsequent scaling applied to a $3 \times 3$ model with a filter of the form

$$\begin{bmatrix} b_0 & b_1 & b_0 \\ b_1 & 0 & b_1 \\ b_0 & b_1 & b_0 \end{bmatrix}. \tag{6.17}$$

We find that Eq. (6.6), $b_0 = -\frac{1}{4}$ and $b_1 = \frac{1}{2}$, corresponds to an image model with constant correlation of $\varrho = \frac{1}{4}$. The good performance of this filter suggests that this model on average fits many natural images. Also note that the ratio $b_1 : b_0$ (before scaling) in the numerical $3 \times 3$ analysis turned out to match exactly the theoretical ratios in the $2 \times 2$ setup, namely

$$\frac{b_1}{b_0} = -\frac{1}{2\varrho}. \tag{6.18}$$

This indicates—but in no way proves—that Eqs. (6.15) and (6.16) might also apply in more general settings and can be used to adapt predictor coefficients to prior knowledge of the strength of linear correlation.

We do not explore this direction further; instead, we propose an alternative method to build an *adaptive predictor* for a particular image under investigation. The idea is to estimate the optimal predictor coefficients $\hat{b}_j$ from the stego image with a least-squares regression. Ordinary least squares (OLS) is

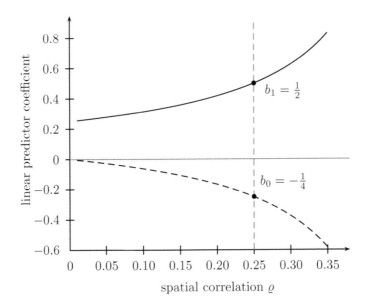

Fig. 6.2: Optimal linear predictor weights for constant spatial correlation model as a function of the pair-wise correlation $\varrho$; filter layout of Eq. (6.17)

appropriate for unweighted WS whereas weighted least squares (WLS) is the method of choice for weighted WS,

$$\hat{b} = \arg\min_{b} \left\| \left( x^{(p)} - \boldsymbol{\Phi} b \right) \odot w \right\| = \left( \boldsymbol{\Phi}^{\mathsf{T}} \operatorname{Diag}(w)\, \boldsymbol{\Phi} \right)^{-1} \boldsymbol{\Phi}^{\mathsf{T}} \operatorname{Diag}(w)\, x^{(p)},$$
(6.19)

with

$$\Phi_{i,1} = x^{(p)}_{\mathsf{Right}(i)} + x^{(p)}_{\mathsf{Left}(i)} + x^{(p)}_{\mathsf{Upper}(i)} + x^{(p)}_{\mathsf{Lower}(i)}$$
(6.20)

and

$$\Phi_{i,0} = x^{(p)}_{\mathsf{Upper}(\mathsf{Right}(i))} + x^{(p)}_{\mathsf{Upper}(\mathsf{Left}(i))} + x^{(p)}_{\mathsf{Lower}(\mathsf{Right}(i))} + x^{(p)}_{\mathsf{Lower}(\mathsf{Left}(i))}.$$
(6.21)

We expect that the so-obtained coefficients also form a good predictor of cover pixels because the LSB replacement process should cause errors of $\pm 1$ in the centre and neighbour pixels approximately equally often. With an adaptive filter the computation time is approximately doubled, as an initial pass through the image is required to determine the filter. A similar technique can be applied to $5 \times 5$ filters, using the symmetrical pattern in (6.22):[2]

$$\begin{bmatrix} b_4 & b_3 & b_2 & b_3 & b_4 \\ b_3 & b_1 & b_0 & b_1 & b_3 \\ b_2 & b_0 & 0 & b_0 & b_2 \\ b_3 & b_1 & b_0 & b_1 & b_3 \\ b_4 & b_3 & b_2 & b_3 & b_4 \end{bmatrix} . \tag{6.22}$$

The performance of our enhanced predictors, both static and adaptive variants, is compared to standard WS and SPA (as representatives of so-called 'structural detectors'; cf. Sect. 2.10.2) below in Sect. 6.1.4.

## 6.1.2 Enhanced Calculation of Weights

In [133], we have proposed an enhanced weighting scheme using Eq. (6.3) with $u = 5$ (instead of $u = 1$ in [73]). This modification was backed with empirical evidence on multiple data sets. And we have speculated that performance gains of so-called *moderated weights* are due to the smaller influence of stego noise, which forms a significant part of the predictor error in areas of low noise. Contrary, standard weights ($u = 1$) appear to emphasise flatter areas in images too much. This explanation is valid; however, it applies only to singular images for which two conditions coincide. In the following, we are able to give a more precise explanation of the advantages of moderated over standard weights. For this purpose, let us decompose the WS estimation equation (2.40) into $k$ layers of similar predictability,[3] $\mathcal{L}_1, \ldots, \mathcal{L}_k$, so that $\mathcal{L}_j = \left\{ i \mid \frac{(j-1)^{\frac{3}{2}}}{4} \leq \mathsf{Conf}(\boldsymbol{x}^{(p)})_i < \frac{j^{\frac{3}{2}}}{4} \right\}$:

$$\hat{p} = 2 \sum_{i=1}^{n} w_i \left( x_i^{(p)} - \overline{x}_i^{(p)} \right) \left( x_i^{(p)} - \mathsf{Pred}(\boldsymbol{x}^{(p)})_i \right) \tag{6.23}$$

$$= 2 \sum_{j=1}^{k} \sum_{i \in \mathcal{L}_j} w_i \left( x_i^{(p)} - \overline{x}_i^{(p)} \right) \left( x_i^{(p)} - \mathsf{Pred}(\boldsymbol{x}^{(p)})_i \right) ; \tag{6.24}$$

and, for unweighted WS or if the layers are homogeneous enough,[4]

---

[2] Note that we changed the enumeration of coefficients. This is intentional as $b_0$ has no special role in the $5 \times 5$ filter anymore.

[3] The predictor was configured as in Eq. (6.6).

$$\approx 2 \sum_{j=1}^{k} \tilde{w}_j \sum_{i \in \mathcal{L}_j} \left( x_i^{(p)} - \overline{x}_i^{(p)} \right) \left( x_i^{(p)} - \mathsf{Pred}(\boldsymbol{x}^{(p)})_i \right). \tag{6.25}$$

Then, for each layer $\mathcal{L}_j$, we compute the MAE of stego pixel prediction,

$$\mathrm{MAE}_{\mathcal{L}_j, \boldsymbol{x}^{(p)}} = \frac{1}{|\mathcal{L}_j|} \sum_{i \in \mathcal{L}_j} \left| x_i^{(p)} - \mathsf{Pred}(\boldsymbol{x}^{(p)})_i \right|, \tag{6.26}$$

the MAE of cover pixel prediction (for analytical reasons only—a real stega-nalyst cannot calculate this measure),

$$\mathrm{MAE}_{\mathcal{L}_j, \boldsymbol{x}^{(0)}} = \frac{1}{|\mathcal{L}_j|} \sum_{i \in \mathcal{L}_j} \left| x_i^{(0)} - \mathsf{Pred}(\boldsymbol{x}^{(p)})_i \right|, \tag{6.27}$$

and lastly the MAE of the (unweighted) secret message length estimation,

$$\mathrm{MAE}_{\mathcal{L}_j, \hat{p}} = \frac{2}{|\mathcal{L}_j|} \sum_{i \in \mathcal{L}_j} \left| \left( x_i^{(p)} - \overline{x}_i^{(p)} \right) \left( x_i^{(p)} - \mathsf{Pred}(\boldsymbol{x}^{(p)})_i \right) \right| = 2 \, \mathrm{MAE}_{\mathcal{L}_j, \boldsymbol{x}^{(p)}}. \tag{6.28}$$

We take averages over $N = 400$ test images (set A, see Appendix A) to allow for meaningful interpretations. This decomposition helps us to study the contribution of individual layers to the total estimation error to gain insight into the 'optimal' weighting scheme.

Note that it is not useful to regard the magnitude of errors in individual layers (nor their sum over multiple layers) since partial MAEs are not additive in the overall estimation equation.[5] However, it is reasonable to assume that the relative contribution of each layer to the overall estimation error is approximately proportional to the error measure of the individual layers.[6] Building on this assumption, Fig. 6.3 plots the approximate cumulative estimation error $\sum_{j=1}^{\iota} \mathrm{MAE}_{\mathcal{L}_j, \hat{p}}$ against the cumulative weight $\sum_{j=1}^{\iota} \tilde{w}_i$ (both scaled to 100%) in the range $\iota = 1, \ldots, k$ for all three weighting schemes:

- standard weights $u = 1$ (dashed line),
- moderated weights $u = 5$ (solid line), and
- unweighted WS $w_i = n^{-1} \Leftrightarrow \tilde{w}_j = |\mathcal{L}_j|/n$ (dotted line).

---

[4] The approximation turns to an identity for unweighted WS.

[5] This is so for two reasons: first, MAEs partly cancel out depending on the covariance of individual (signed) residuals; second, the overall error is expected to be smaller than the errors in individual layers due to the law of large numbers.

[6] This is based on the assumption that, on average over all images, the covariance between errors in individual layers does not vary systematically with the predictability measure used to defined the layers, i.e., function $\mathsf{Conf}(\boldsymbol{x}^{(p)})$.

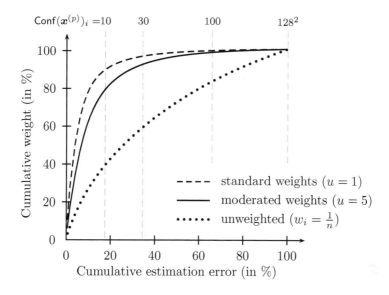

Fig. 6.3: Cumulative mean absolute estimation error (MAE) over layers of equi-predictability against cumulative weight of aggregated layers (average over $N = 400$ images $640 \times 480$, $p = 0$)

Selected values of the predictability measure (by which layers $\mathcal{L}$ are ordered) are annotated on the horizontal axis. Several observations from the figure are noteworthy:

1. Even in the unweighted case, the relative contribution is skewed to the better-predictable layers. This is a result of spatial correlation in typical images, i.e., smooth (and thus more predictable) areas are over-represented on average.
2. Standard weighting shifts large parts of the total weight to better predictable (lower MAE) layers. This explains the improved performance of weighted WS over unweighted WS.
3. Moderated weighting, however, pushes back the distribution of weight somewhat to the unweighted case. So, on average, larger errors in higher-order layers $\mathcal{L}_j$ are more likely to disturb the overall secret message length estimate. Hence, Fig. 6.3 alone cannot explain why moderated weights empirically lead to *lower* errors, and leaves us puzzled in this regard.

Note that the curve in Fig. 6.3 does not change substantially for $p > 0$, even if $\mathrm{MAE}_{\mathcal{L}_j, \hat{p}}$ of Eq. (6.28) is replaced with $\mathrm{MAE}_{\mathcal{L}_j, \boldsymbol{x}^{(0)}}$, Eq. (6.27), as error measure. We omit these curves, as the differences are barely informative (and cannot explain the merits of moderated weights either).

The only conclusion we can draw from the above analysis (and many variants thereof not reported here) is that moderated weights are not superior to standard weights across all images equally, but the experimentally confirmed performance gains of moderated weights might be caused by hidden hetero-geneity: aggregate performance measures are driven by a few images in the data set, for which standard weights produce serious outliers.

From Eq. (2.45) we know that flat areas, which often concur with satu-ration, lead to bias in the estimator. More precisely, *parity co-occurence*, the fact that a centre pixel and most of its neighbours share the same parity, is the root cause of this bias (flat areas are a sufficient but not necessary condition for parity co-occurence). The bias is caused by individual errors taking the same sign, so that they sum up instead of cancelling out in the es-timation equation. As parity co-occurrence is a rare phenomenon in our test data (cf. Tab. A.1 in Appendix A), and WS can cope well if it appears only locally, parity co-occurrence must stand in an even more complicated relation to the weighting scheme: images become outliers under standard weights *only if* parity co-occurrence exists *and* there are no sufficient smooth gradients in the image to compensate it.

Let us discuss this finding on the example images shown in Fig. 6.4. Both images exhibit some flat areas due to saturation, yet only the left image (with a smaller share of saturated pixels) leads to serious bias in weighted WS. The reason for this becomes evident from Fig. 6.5, where cardinalities of $|\mathcal{L}_j|$ are plotted in histogram style, with (unscaled) weights as a function of $\mathsf{Conf}(\boldsymbol{x}^{(p)})$ superimposed. Observe the 'gap' in the histogram of Fig. 6.5 (a). After normalisation, it shifts a substantial mass of weight to the flat areas (where $\mathsf{Conf}(\boldsymbol{x}^{(p)}) = 0$), so that bias accumulated there kicks through to the estimate $\hat{p}$. The soft gradients on the wall of the house fill this gap for the right image (Fig. 6.5 (b)) and attain enough weight to offset, after normalisation, the bias due to the flat areas. Moderated weights remedy this discrepancy somewhat by setting a lower starting point at the left end of the weight distribution function. This ensures that higher-order layers $\mathcal{L}_j$ attain enough weight to prevent serious outlier estimates $\hat{p}$ (though still far apart from the true embedding rate $p$). So, effectively, moderated weights are a compromise between weighted and unweighted WS. They combine the higher accuracy of the former while mitigating the risk of outliers somewhat. The overall effect is that moderated weights score better on performance measures that are aggregated over heterogeneous image sets.

In fact, this finding suggests that the theory of heterogeneous covers could be taken down one level to explain differences in areas of covers, e.g., by using the mixture cover model on the level of individual samples. This can be linked to the idea of segmentation mentioned in [118] and to the concept of superposition models for images in [221]. Further refinements based on this approach, however, are beyond the scope of this book.

(a) $\hat{p}_{\mathrm{unw.}}$=0.01, $\hat{p}_{u=1}$=0.55, $\hat{p}_{u=5}$=0.29    (b) $\hat{p}_{\mathrm{unw.}}$=0.06, $\hat{p}_{u=1}$=0.02, $\hat{p}_{u=5}$=0.01

Fig. 6.4: Example eight-bit greyscale images taken from a digital camera; both images contain saturated areas; however, only the left image causes outliers in weighted WS analysis; estimates given for $p = 0$; image source: set A

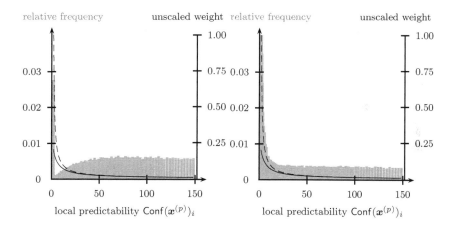

(a) no unsaturated smooth areas    (b) WS works despite saturation

Fig. 6.5: Histograms of predictability measure for the examples images in Figure 6.4 (left-hand scale) and unscaled weights $w_i$ as a function of $\mathsf{Conf}(\boldsymbol{x}^{(p)})_i$; standard weights ($u = 1$, dashed line) and moderated weights ($u = 5$, solid line; both right-hand scale)

### 6.1.3 Enhanced Bias Correction

Another way to deal with parity co-occurrence is compensation by bias correction. Our proposed correction term can be explained by the expected value of the term annotated with 'predicted stego noise' in the error decomposition of Eq. (2.45),

$$\mathsf{E}\left(\hat{p} - p\right) = \mathsf{E}\left(\left(\mathsf{Pred}(\boldsymbol{x}^{(0)})_i - \mathsf{Pred}(\boldsymbol{x}^{(p)})_i\right)\left(x_i^{(p)} - \overline{x}_i^{(p)}\right)\right). \tag{6.29}$$

Assuming a linear filter for function $\mathsf{Pred}$, we can rewrite the expression as

$$\mathsf{E}\left(\hat{p} - p\right) = \mathsf{E}\left(\left(\mathsf{Pred}(\boldsymbol{x}^{(0)} - \boldsymbol{x}^{(p)})_i\right)\left(x_i^{(p)} - \overline{x}_i^{(p)}\right)\right). \tag{6.30}$$

The expected error is only zero if the filtered added stego signal $\boldsymbol{x}^{(p)} - \boldsymbol{x}^{(0)}$ is independent of the corresponding centre pixel. This is false if there is parity co-occurrence between neighbours in the cover for the reasons given in Sect. 2.10.4 (p. 75).

To quantify this bias, we imagine that it is the *stego* image which is fixed and the *cover* which was generated by randomly flipping proportion $p/2$ of LSBs (this disregards certain conditional probabilities in the structure of the cover, but that is not very significant). Then we can simplify the expected bias, the last part of (2.45), to

$$\hat{d} = 2\sum_{i=1}^{n} w_i \left(x_i^{(p)} - \overline{x}_i^{(p)}\right) \mathsf{E}\left(\mathsf{Pred}\left(\boldsymbol{x}^{(0)} - \boldsymbol{x}^{(p)}\right)_i\right) \tag{6.31}$$

$$= \hat{p}\sum_{i=1}^{n} w_i \left(x_i^{(p)} - \overline{x}_i^{(p)}\right) \left(\mathsf{Pred}\left(\overline{\boldsymbol{x}}^{(p)} - \boldsymbol{x}^{(p)}\right)_i\right) \tag{6.32}$$

since the filter is linear, and $(\boldsymbol{x}^{(0)} - \boldsymbol{x}^{(p)})_i$ is $(\overline{\boldsymbol{x}}^{(p)} - \boldsymbol{x}^{(p)})_i$ with probability $p/2$ and zero otherwise. Given an initial estimate of $\hat{p}$, one can subtract the expected bias $\hat{d}$ to make the estimate more accurate. The results in the next section show that this makes a substantial improvement to the accuracy of the estimator in covers where there is strong parity co-occurrence between neighbouring pixels in the cover image.

### 6.1.4 Experimental Results

The objective of this section is to validate the performance increases due to the proposed enhancements on image sets with controlled heterogeneity. The individual improvements of the WS method presented so far define a design space of 45 different detector variants. Since a comparison of all variants on reasonably large and diverse image sets is computationally expensive and difficult to present concisely and readably, we focus our study on the most relevant options as indicated in Table 6.1. In particular, we refrain from reporting results for standard bias correction of Eq. (2.45) as well as for the fixed coefficient $3 \times 3$ predictor of Eq. (6.5), because they are clearly inferior compared to other options. We further omit for brevity the results for the $3 \times 3$ adaptive predictor and do not benchmark enhanced bias correction (Eq. (6.31)) for other combinations than the best performing $5 \times 5$ predictor. As the emphasis in this book is on explaining *why* certain improvements work (e.g., predictor coefficients, Sect. 6.1.1; moderated weights, Sect. 6.1.2), we refer the reader to our published results in [133] for a more comprehensive analysis with regard to the partial contribution of the size of the filter window and combinations of adaptive coefficients and bias correction.

Table 6.1: Options for WS analysis considered in the experimental evaluation

| Predictor | Weighting scheme | | | | | | | | |
|---|---|---|---|---|---|---|---|---|---|
| | unweighted | | | standard ($u = 1$) | | | moderated ($u = 5$) | | |
| *bias correction:* | none | std. | enh. | none | std. | enh. | none | std. | enh. |
| standard, Eq. (6.4) | × | | | × | | | | | |
| standard, Eq. (6.5) | | | | | | | | | |
| enhanced $3 \times 3$ (6.6) | × | | | × | | | × | | |
| adaptive $3 \times 3$ (6.17) | | | | | | | | | |
| adaptive $5 \times 5$ (6.22) | × | | × | × | | | × | | × |

Combinations marked with '×' are included in the experimental validation exercise.

The reference set for the experiments in this section are $N = 1,600$ images of image set A (cf. Appendix A). All images were obtained from a digital camera in raw format. Every image was reduced from $2,000 \times 1,500$ pixels to $640 \times 480$ pixels using three interpolation methods and, for a fourth comparison group that preserved the original dependencies between pixels, by random cropping. The rationale for using different downsampling methods here is that WS crucially depends on the accuracy of the local predictor, which might be affected by the type and amount of local correlation introduced by interpolation algorithms used for downsampling [225], or, in the case of the cropped images, by *colour filter array* (CFA) demosaicking inside the camera [200]. The reduced-size images were converted to eight-bit grey scale. We measure the performance of all WS variants under investigation plus SPA, using all overlapping vertical and horizontal pairs (Sect. 2.10.2,

[50]),[7] by the MAE,

$$\mathrm{MAE}(p) = \frac{1}{N} \sum_{j=1}^{N} |\hat{p}_j - p|$$

$$= \frac{1}{N} \sum_{j=1}^{N} \left| \mathsf{Detect}_{\mathsf{Quant}} \left( \mathsf{Embed} \left( \{0,1\}^{pn}, \boldsymbol{x}_j^{(0)}, \boldsymbol{k} \right) \right) - p \right|, \quad (6.33)$$

a robust compound error measure,[8] as a function of the net embedding rate $p$. We do not separate error sources by iterating the secret message for fixed $j$ and $p$ [22].

The baseline results in Fig. 6.6 compare the variant of enhanced WS with a $5 \times 5$ adaptive predictor (Eqs. (6.19) and (6.22)), moderated weights (Eq. (6.2) with $u = 5$) and enhanced bias correction (Eq. (6.32)), which we consider as the default for enhanced WS, to the previously known variants of WS and SPA. It is clearly visible from all curves that enhanced WS on average produces the lowest estimation errors of all compared methods independently of the embedding rate $p$ and the type of covers. At the same time, it becomes apparent that differences in detectability persist between cover sets. The comparative advantage of enhanced WS over SPA is highest for cropped images, followed by bicubic downsampling. Presumably this is due to higher spatial correlation in these image sets compared to edgier downsampling methods, such as bilinear or nearest neighbour interpolation. The latter method generally leads to hard-to-steganalyse covers throughout all detectors. This performance difference between cover sets once again highlights the effects of heterogeneous covers and its associated risk of false generalisations from those of studies that fail to deal with heterogeneity appropriately.

Figure 6.7 breaks down the performance differentials for individual components of the enhanced WS method, namely predictor type and weighting scheme. Observe that the difference between the $3 \times 3$ enhanced predictor (grey curves) and the $5 \times 5$ adaptive predictor (black curves) is rather small,[9] and of some significance only for cropped images. This can be interpreted as an indication that a larger adaptive filter might be better suited to capture camera-specific spatial correlation while at the same time being comparatively less sensitive to spatial noise than smaller filters. The $5 \times 5$

---

[7] Discrepancies between our results for SPA at embedding rates $p > 0.7$ and those reported in the literature are due to the fact that we ignore cases where SPA 'fails' (i.e., the estimation equation (2.34) has no real roots) in the computation of performance indicators, while some authors set $\hat{p} = 1$ in this case. There is no optimal method to deal with detector failures in aggregate indicators, but we believe that omission is the more natural choice. See Fig. G.4 in Appendix G.

[8] 'Robust' refers to the moderate sensitivity to individual outliers; 'compound' means that both within-image and between-image error are measured together.

[9] The $3 \times 3$ adaptive is somewhere between the two, which is why we decided not to show the results.

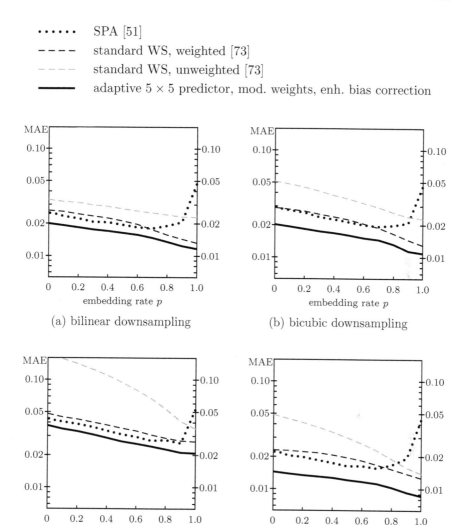

Fig. 6.6: Baseline results for performance of enhanced WS: Mean absolute estimation error as a function of the embedding rate $p$ for different types of covers; smaller numbers indicate better performance; $N = 1,600$ never-compressed camera images reduced to $640 \times 480$; note the log scales

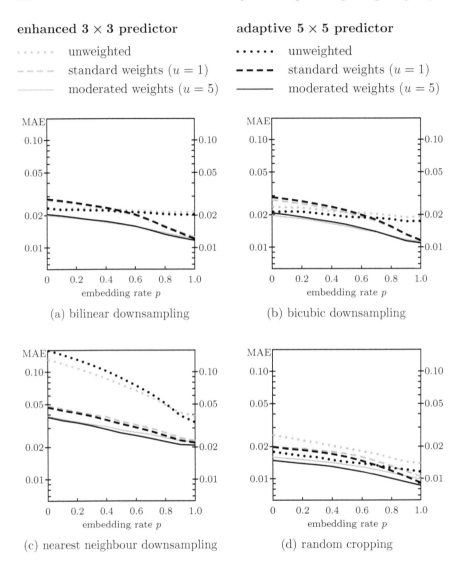

Fig. 6.7: Performance gains of selected individual enhancements to components of the WS method: MAE as a function of the embedding rate $p$ for different types of covers; smaller numbers indicate better performance; $N = 1,600$ never-compressed camera images reduced to $640 \times 480$; note the log scale

adaptive filter performs equally or slightly better, with a single exception for unweighted WS on nearest neighbour downsampled covers at high embedding rates $p > 0.8$.

In general, the weighting scheme is much more influential than the choice of the predictor. Using weights at all is crucial for the 'nearest neighbour' set, and somewhat beneficial in all other cases. The advantage of moderated weights ($u = 5$) over standard weights ($u = 1$) is visible throughout all sets, as some images with parity co-occurrence and a 'gap' in the predictability histogram (cf. Fig. 6.5 (a), Sect. 6.1.2) can be found in each of them. Observe that after downsampling with bilinear or bicubic interpolation, unweighted variants of enhanced WS perform better than standard weights.

The aggregated figures hide the heterogeneity between images, which explains the average performance differential. So the *individual* analysis in Fig. 6.8, based on all covers ($p = 0$) downsampled with bicubic interpolation, is more telling. Figure 6.8 (a) displays the absolute estimation error $|\hat{p}|$ as a function of the (normalised) weight assigned to flat areas (i.e., $\mathsf{Conf}(x^{(0)})_i = 0$) for standard ($u = 1$) and moderated ($u = 5$) weights. Arrows connect data points belonging to the same cover image. Under both weighting schemes, the positive relation between the high weight of flat areas and large estimation errors is very well visible.

Observe that the high weight of flat areas is a necessary but not sufficient condition for bad detection performance. This confirms our argument in Sect. 6.1.2 that high parity co-occurrence can be compensated for in some images if smooth and thus well-predictable areas coexist, as for instance in Fig. 6.4 (b). Overall, as moderated weights reduce the fraction of weight assigned to flat areas, the aggregate estimation error decreases, though only somewhat for the worst-affected images and yet apparently enough to bring down the aggregate measure. This is so because the sensitive images constitute only a small fraction of all images in our data set (see Fig. 6.8 (b) for a 'delta view' on the data points of Fig. 6.8 (a) on a different scale). So, for our data set, $u = 5$ seems to be a good compromise, but we cannot generalise this finding unless we repeat the analysis on many (representative) data sets.

Nevertheless, as a robustness check, the same analysis as in Fig. 6.7 has been repeated on image set B, all never-compressed scans, and the results are reported in Fig. G.3 of Appendix G. The relative performance is broadly similar, with the main differences being a generally larger disadvantage of unweighted variants and an inverse-U relation between MAE and embedding rate $p$, which is mainly caused by bias due to parity co-occurrence.

This leads us to the effect of enhanced bias correction. We only report results for two sets of downsampled (bicubic) images in Fig. 6.9, as the curves for all other sets are not substantially different. It is striking that the influence of bias correction is hardly measurable for image set A (regardless of the weighting scheme). This is probably due to the generally low level of parity co-occurrence in this data set, a conjecture that can be supported with the following analysis on image set B: here, bias correction helps a lot for $p < 0.8$,

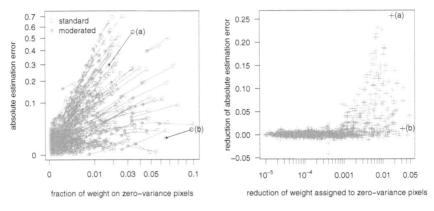

(a) weighting and performance     (b) standard minus moderated weights

Fig. 6.8: Effect of weighting schemes on heterogeneous images; $N = 1{,}600$ never-compressed camera images downsampled (bicubic) to $640 \times 480$; $\mathsf{Conf}(\boldsymbol{x}^{(0)})$ and $\hat{p}$ measured on plain covers ($p = 0$); highlighted data points refer to the example images displayed in Fig. 6.4; note the scales

but the effect vanishes when we exclude 20% of the images with the highest level of parity co-occurrence (dashed curves in Fig. 6.9 (b)). The fact that the inverse-U shape attenuates after exclusion also supports our argument that this shape is mainly caused by estimation bias. Further results in [133] suggest that our enhanced bias correction also helps us to cope with other sources of parity co-occurrence beyond saturation (as is predominately the case for images of set B). Such sources include artificial denoising inside digital cameras and related post-processing.

All in all, we could validate with experiments that the proposed enhancements to WS in fact improve detection performance and make enhanced WS the preferred method for quantitative steganalysis of LSB replacement in never-compressed covers. This view is also confirmed by additional performance indicators, including measures of bias and dispersion, which are reported in Table G.9 of Appendix G. The next section deals with a special case which has been disregarded so far, namely when covers are not just down-scaled but decompressed from lossy compression, such as JPEG.

## 6.2 Adaptation of WS to JPEG Pre-Compressed Covers

The results reported in [133] and in the previous section allow to conclude that for never-compressed covers, the performance of enhanced WS is (almost

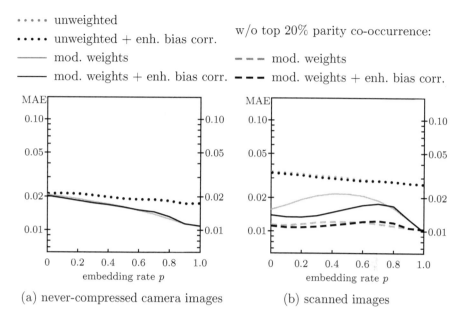

Fig. 6.9: Effect of improved bias correction for WS analysis: MAE as a function of the embedding rate $p$ for different cover sources; adaptive $5 \times 5$ predictor; smaller numbers indicate better performance; $N = 1{,}600$ never-compressed camera images downsampled (bicubic) to $640 \times 480$ (left) and $N = 2{,}945$ scanned images downsampled (bicubic) to $640 \times 457$ (right); note the log scale

always) superior to that of any other detector, while its computational complexity is comparatively low. However, enhanced WS still does not match the performance of structural detectors (see Sects. 2.10.2 and 2.10.3) if the cover image has been compressed with JPEG at some stage before embedding [22]. This is a substantial drawback because many image acquisition devices, foremost digital cameras in the consumer segment, store JPEG images by default. To close this gap, a modification of WS optimised for JPEG covers is presented and benchmarked in this section.

### 6.2.1 Improved Predictor

We now describe a new local predictor $\mathsf{Pred_{JPEG}}$ targeted to JPEG pre-compressed covers, i.e., the spatial domain image representation of the cover

$\boldsymbol{x}^{(0)}$ is a result of a JPEG decompression operation. Our predictor works on former JPEG blocks and exploits the constraints imposed on possible realisations in $\mathbb{Z}^{64}$ through quantisation with factors greater than 1:

$$\mathsf{Pred}_{\mathsf{JPEG}}\left(\boldsymbol{x}_{\boxplus}^{(p)},\hat{q}\right) = \boldsymbol{a}_{2\mathrm{D}}^{\mathsf{T}}\,\boldsymbol{q}\,\left\lfloor\overline{\boldsymbol{q}}\,\boldsymbol{a}_{2\mathrm{D}}\,\boldsymbol{x}_{\boxplus}^{(p)} + \frac{1}{2}\right\rfloor. \tag{6.34}$$

Matrix $\boldsymbol{a}_{2\mathrm{D}}$ is the 2D-DCT transformation matrix as in Eq. (2.5) and diagonal matrices $\boldsymbol{q}$, $\overline{\boldsymbol{q}}$ hold the (inverse) quantisation factors as defined in Eq. (2.7). Scalar $\hat{q}$ is the (estimated) JPEG quality of the preprocessing operation.

   To show why and under which conditions Eq. (6.34) is a good predictor, let us decompose the image under analysis as the cover plus additive stego noise,

$$\boldsymbol{x}_{\boxplus}^{(p)} = \boldsymbol{x}_{\boxplus}^{(0)} + \overbrace{\left(\boldsymbol{x}_{\boxplus}^{(p)} - \boldsymbol{x}_{\boxplus}^{(0)}\right)}^{\text{stego noise}} = \boldsymbol{a}_{2\mathrm{D}}^{\mathsf{T}}\,\boldsymbol{q}\,\boldsymbol{y}_{\boxplus}^{(0)*} + \boldsymbol{\epsilon} + \left(\boldsymbol{x}_{\boxplus}^{(p)} - \boldsymbol{x}_{\boxplus}^{(0)}\right), \tag{6.35}$$

where $\boldsymbol{y}_{\boxplus}^{(0)*}$ is the matrix of quantised cover DCT coefficients that appears when we rewrite cover $\boldsymbol{x}^{(0)}$ as a result from a JPEG decompression operation. Vector $\boldsymbol{\epsilon}$ is the additive rounding error due to the casting of real intensity values to integers in the spatial domain. By pre-multiplying the joint error term (sum of stego noise and rounding error) with the orthogonal transformation matrix and rearranging, we obtain

$$\boldsymbol{x}_{\boxplus}^{(p)} = \boldsymbol{a}_{2\mathrm{D}}^{\mathsf{T}}\left(\boldsymbol{q}\,\boldsymbol{y}_{\boxplus}^{(0)*} + \underbrace{\boldsymbol{a}_{2\mathrm{D}}\left(\boldsymbol{\epsilon} + \boldsymbol{x}_{\boxplus}^{(p)} - \boldsymbol{x}_{\boxplus}^{(0)}\right)}_{\text{DCT trans. of joint error}}\right) = \boldsymbol{a}_{2\mathrm{D}}^{\mathsf{T}}\left(\boldsymbol{q}\,\boldsymbol{y}_{\boxplus}^{(0)*} + \tilde{\boldsymbol{y}}_{\boxplus}^{(p)}\right).$$
$$\tag{6.36}$$

Equation (6.34) is a good predictor for $\boldsymbol{x}^{(0)}$ on average if the DCT transformation of the joint error $\tilde{\boldsymbol{y}}_{\boxplus}^{(p)}$ is below half of one quantisation factor. In this case, the requantisation in the DCT domain can undo the embedding changes and revert to the cover pixels.

   We can approximate the probability that the joint error remains below the quantisation margin for each subband by modelling the error term as a sum of independent random variables with known variance. The pixel rounding error is assumed to follow a uniform distribution in the interval $[-0.5, +0.5)$.[10] The stego signal is a random variable with three possible realisations, 0 (with probability $(1 - p/2)$), $-1$, and $+1$ (each with probability $p/4$). Note that we can ignore the structural dependence typical for LSB replacement between the sign of the stego signal and the parity of the cover. The distribution of

---

[10] This assumption may be violated for saturated pixels, which need special treatment in WS. Our test images in set A exhibit comparatively less-saturated areas.

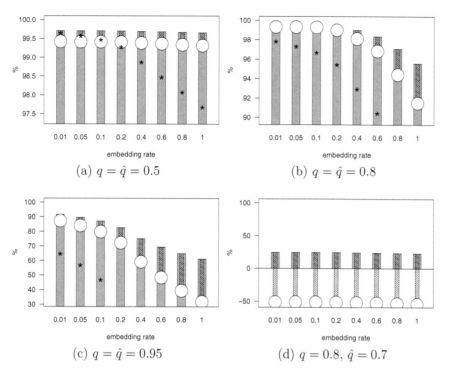

Fig. 6.10: JPEG predictor accuracy: percentage of embedding changes reverted to cover values after requantisation (grey bars) and theoretical bound (Eq. (6.39), black asterisks), percentage of non-embedding positions altered after requantisation (hatched bars), total accuracy measured as difference between the two (white disks); note the different scales

the sum of pixel rounding error and stego signal has mean 0 and variance $\frac{1+6p}{12}$. Each coefficient in $\tilde{\boldsymbol{y}}_{\boxplus}^{(p)}$ of the joint error is the weighted sum of 64 uncorrelated realisations, so the variance of the compound distribution in the transformed domain is

$$\tilde{\sigma}^2 = (a_{2\mathrm{D}_{i,1}})^2 \frac{1+6p}{12} + \cdots + (a_{2\mathrm{D}_{i,64}})^2 \frac{1+6p}{12} = \frac{1+6p}{12} \sum_{j=1}^{64} (a_{2\mathrm{D}_{i,j}})^2 = \frac{1+6p}{12}.$$

$$(6.37)$$

The last identity follows from the orthogonality of $\boldsymbol{a}_{2\mathrm{D}}$: squared elements add up to 1 row- and column-wise. We can obtain a lower bound for the probability that coefficients in $\tilde{\boldsymbol{y}}_{\boxplus}^{(p)}$ are below half of the quantisation factor from Chebychev's inequality:

$$\text{Prob}\left(\left|\tilde{\boldsymbol{y}}_{\boxplus i,j}^{(p)}\right| \geq k\right) \leq \frac{\tilde{\sigma}^2}{k^2} \quad \text{with} \quad k = \frac{\text{Quant}(q,i)}{2}, \tag{6.38}$$

$$\text{Prob}\left(\left|\tilde{\boldsymbol{y}}_{\boxplus i,j}^{(p)}\right| < \frac{\text{Quant}(q,i)}{2}\right) \geq 1 - \frac{4\,\tilde{\sigma}^2}{\text{Quant}(q,i)^2} = 1 - \frac{(1+6p)}{3\,\text{Quant}(q,i)^2}. \tag{6.39}$$

For example, the smallest[11] quantisation factor at $q = 0.8$ is 4, so that at least 97% of the cover pixels can be correctly predicted for embedding rates up to $p = 0.05$, 92% for $p = 0.5$, and 85% for maximum capacity $p = 1$. Note that these estimates are very conservative, because Eq. (6.39) is a rather loose bound and quantisation factors for higher-order subbands are up to ten times larger than the infimum. In practice, the chances of predicting cover pixels correctly are even higher, as can be seen from the results of experiments on real images displayed in Fig. 6.10 (a)–(c). The bars decompose overall predictor accuracy by the percentage of embedding changes revertible to cover values after requantisation (grey bars), from which the percentage of non-embedding position inadvertently modified has to be deducted (hatched bars). Observe that the predictor accuracy depends on both $p$ and $q$ broadly in the theoretically predicted way.

It is difficult to derive analytical results for the accuracy of Eq. (6.34) if $2\left|\tilde{\boldsymbol{y}}_{\boxplus i,j}\right| > \text{Quant}(q,i)$ for one or more coefficients in a block, because hardly tractable interdependencies between pixels emerge. Even if the predictor properties do not degrade severely on average, we have to raise an important assumption of the WS method, namely that the prediction error $\text{Pred}_{\text{JPEG}}\left(\boldsymbol{x}^{(p)}\right)_i - x_i^{(p)}$ is independent of the stego noise at position $i$. Another source of hard-to-control deviations from the theoretical results could be error correlation due to propagation in efficient DCT implementations, such as FDCT. We resort to an experimental robustness test to gauge the influence of the DCT algorithm.

## 6.2.2 Estimation of the Cover's JPEG Compression Quality

The local predictor $\text{Pred}_{\text{JPEG}}$ in (6.34) depends on $\hat{q}$, the quality factor of the JPEG pre-compression, which is unobservable to the steganalyst and has to be estimated. Knowledge of $\hat{q}$ is crucial: as shown in Fig. 6.10 (d), the capability of reverting stego changes breaks down if the estimate $\hat{q}$ does not match the true compression quality $q$.

There exists some literature on JPEG quantisation matrix estimation. One option is the method described in the appendix of [75], which estimates

---

[11] We use the smallest quantisation factor since deviations beyond one rounding margin in a single coefficient may affect others in a nontrivial way.

*individual* quantisation factors from the distribution of DCT coefficients calculated from the image under investigation. Our approach is closer to the method of Fan and Queiroz [55], who maximise a likelihood function over the space of *standard matrices*, a particular choice of function Quant used in the libjpeg reference implementation [111]. While their method focuses on *detecting the fact* of JPEG pre-compression, and its validation is limited to ten test images without providing reliable figures on the accuracy of quality *estimates*, Pevný and Fridrich [188] employ machine learning techniques to tackle the harder problem of estimating the primary quantisation matrix in double-compressed images, both before and after stego noise has been added.

Similarly to [55], we estimate the *entire matrix* of quantisation factors via the common quality parameter $q$. Our method searches the first local minimum in a series of MSEs calculated in the spatial domain between the original image and re-compressed versions with increasing quality parameters $q$. We acknowledge that our approach is less robust against irregular or unknown functions Quant than [75] and probably [188], but presumably more reliable if Quant is known (as is the case for the lion's share of JPEG images in circulation). It also proved more robust than our implementation of the method in [75] when large parts of the image under investigation have been modified with LSB replacement after decompression. Table 6.2 summarises the success rate (i.e., $\hat{q} = q$) for different qualities $q$ and embedding rates $p$. Observe that our method fails in precisely those cases where JPEG WS is not the optimal detector anyway (see below in Sect. 6.2.3). Since wrong estimates of $\hat{q}$ appear as never-compressed, this method is a safe decision criterion for the subsequent steganalysis method, namely enhanced WS, for seemingly never-compressed images and JPEG WS if $\hat{q}$ exists.

We acknowledge that it is possible that our laboratory results are a bit on the optimistic side, mainly due to the large step sizes used in the evaluation. Therefore, the issue of JPEG quality estimation might deserve more attention in practice. Nevertheless, since machine learning techniques have been successfully used to quite accurately estimate even non-standard primary quantisation matrices after a secondary compression [161] and superposition with stego-noise [189], we are quite optimistic that this detail is resolvable.

## 6.2.3 Experimental Results

The reference set for the experiments in this section are $N = 1,600$ images of image set A (cf. Appendix A). All images were obtained from a digital camera in raw format, downsampled to $640 \times 480$ using Photoshop's bilinear resampling method and then converted to eight-bit grey scale. Although downsampling alters local statistics and might affect steganalysis results in general (see Sect. 6.1.4 and [15]), we conjecture that the results for JPEG WS are less sensitive to such influence because compression artefacts largely

Table 6.2: Success rate for correct estimation of pre-compression quality $\hat{q} = q$

| embedding | never- | JPEG covers with quality $q$ | | | | | | |
|---|---|---|---|---|---|---|---|---|
| rate $p$ | compressed | 0.5 | 0.6 | 0.7 | 0.8 | 0.9 | 0.95 | 0.99 |
| 0.00 | 100.0 | 100.0 | 100.0 | 100.0 | 100.0 | 100.0 | 100.0 | 0.0 |
| 0.01 | 100.0 | 100.0 | 100.0 | 100.0 | 100.0 | 100.0 | 100.0 | 0.0 |
| 0.05 | 100.0 | 100.0 | 100.0 | 100.0 | 100.0 | 100.0 | 100.0 | 0.0 |
| 0.10 | 100.0 | 100.0 | 100.0 | 100.0 | 100.0 | 100.0 | 89.9 | 0.0 |
| 0.20 | 100.0 | 100.0 | 100.0 | 100.0 | 100.0 | 100.0 | 0.0 | 0.0 |
| 0.40 | 100.0 | 100.0 | 100.0 | 100.0 | 100.0 | 100.0 | 0.0 | 0.0 |
| 0.60 | 100.0 | 100.0 | 100.0 | 100.0 | 100.0 | 100.0 | 0.0 | 0.0 |
| 0.80 | 100.0 | 100.0 | 100.0 | 100.0 | 100.0 | 96.6 | 0.0 | 0.0 |
| 1.00 | 100.0 | 100.0 | 100.0 | 100.0 | 99.2 | 51.7 | 0.0 | 0.0 |

All misclassified images were mistaken for never-compressed covers; $N = 385$ images.

erase the more subtle traces of resampling. This is why we deem it more informative to break down the results by different JPEG compression rates rather than test more than one downsampling method.

The performance of WS with adopted predictor for JPEG covers (6.34), weighted and unweighted, has been benchmarked against standard WS (Sect. 2.10.4, [73]), enhanced WS with a $5 \times 5$ adaptive filter (Sect. 6.1), and SPA using all overlapping vertical and horizontal pairs (Sect. 2.10.2, [50]). SPA is a representative of structural detectors known for its robustness against cover pre-compression [22]. The predictor accuracy, measured by the MAE (Eq. (6.33)) is reported in Fig. 6.11. For moderate compression qualities, as expected, SPA is largely more accurate than the established variants of WS (standard and enhanced). The proposed JPEG WS method, however, has an additional advantage over SPA, by up to one order of magnitude for $q = 0.5$. This advantage decreases gradually for higher pre-compression qualities and more so for large embedding rates. Note that weighting does not improve the accuracy of JPEG WS, which is probably due to the standard function Conf being a bad measure of predictability in entire JPEG blocks. Unweighted JPEG WS turns out to be the most accurate quantitative detector for pre-compressed covers up to JPEG qualities $q = 0.8$. Above this threshold it remains, by a large margin, the best discriminator between stego objects and plain covers (see ROC curves in Fig. 6.12). Additional performance indicators can be found in Table G.10 of Appendix G.

We have also tested the reliability of JPEG WS in the hypothetical case when $\hat{q}$ is estimated close but not equal to actual $q$. The results can be found in Fig. 6.13. For low qualities ($q = 0.5$), a discrepancy of $\pm 1$ percentage point is tolerable, although between 45 and 54 quantisation factors do change by at least one step. Larger deviations are generally penalised with a drop in

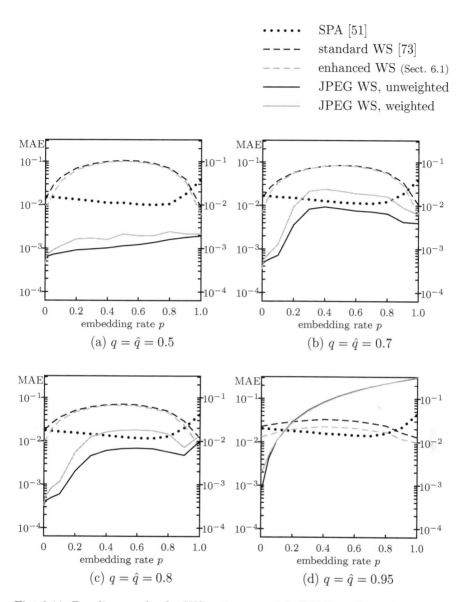

Fig. 6.11: Baseline results for WS estimator with JPEG predictor $\mathsf{Pred_{JPEG}}$: Mean absolute estimation error as a function of the embedding rate $p$ for different JPEG qualities $q$ and perfect quality estimation, i.e., $\hat{q} = q$; smaller numbers indicate better performance; $N = 1,600$ JPEG pre-compressed images (set A); note the log scale

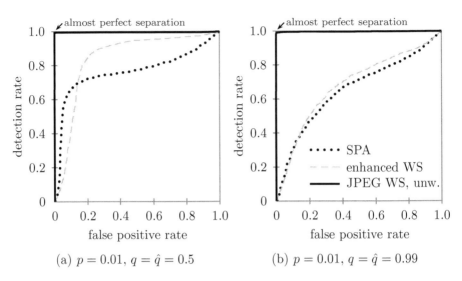

(a) $p = 0.01$, $q = \hat{q} = 0.5$           (b) $p = 0.01$, $q = \hat{q} = 0.99$

Fig. 6.12: ROC curves: JPEG WS has unmatched discriminatory power for low embedding rates even at high quality factors $q$; $N = 1,600$ JPEG pre-compressed images

performance. With increasing $q$, it becomes more important to get $\hat{q}$ right. However, when in doubt between two subsequent values, a slight underestimation of $\hat{q}$ apparently retains higher performance than overestimation.

Differences in the JPEG DCT algorithms used for pre-compression and in the JPEG WS predictor pose another source of error. To explore the robustness of our method to variations of the DCT implementation, we have conducted experiments in which we systematically varied the DCT algorithms at both stages in the three methods offered by the `libjpeg` [111] library, namely 'float', 'fast', and the default value 'slow'. Figure 6.14 shows the largest measurable deviations between combinations of pre-compression and estimation DCT algorithms for qualities $q = \hat{q} = 0.8$ and $q = \hat{q} = 0.95$. Observe that the differences are hardly measurable; the sloppy 'fast' DCT algorithm merely performs a little worse, regardless of the pre-compression algorithm. We therefore do not recommend this performance-optimised DCT method for the JPEG predictor in WS analysis. Note that we have also generated data for $q = \hat{q} = 0.5$ but could not find any measurable influence of the DCT method. The same applies to the combinations of DCT algorithms not shown in the figure. So as far as the tested algorithms are concerned, we may conjecture that our improved WS detector works quite robustly on covers from unknown pre-compression DCT algorithms. This is an important prerequisite for practical steganalysis, where prior knowledge about the compression algorithm is generally not available, and we are not aware of methods to estimate the DCT algorithm from artefacts in the data.

(a) $q = 0.5$ (b) $q = 0.7$

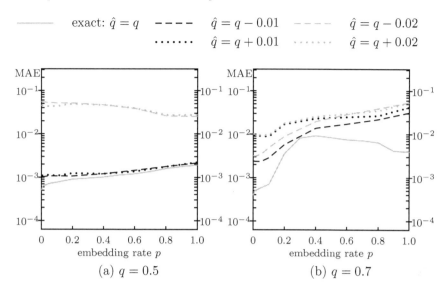

Fig. 6.13: Sensitivity of JPEG predictor $\mathsf{Pred}_{\mathsf{JPEG}}$ to small deviations between the predictor parameter $\hat{q}$ and the actual pre-compression quality $q$; unweighted JPEG WS for all curves; lower MAEs indicate better performance; $N = 400$ JPEG pre-compressed images; note the log scale

(a) $q = \hat{q} = 0.8$ (b) $q = \hat{q} = 0.95$

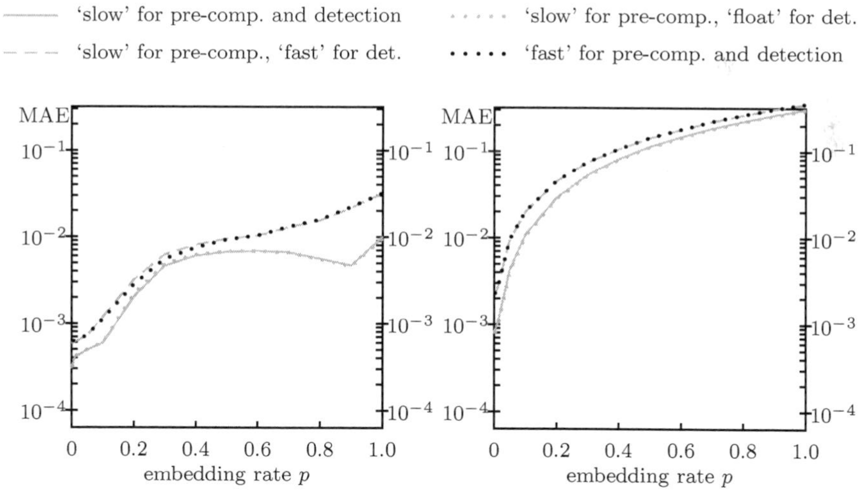

Fig. 6.14: Sensitivity of JPEG predictor $\mathsf{Pred}_{\mathsf{JPEG}}$ to variations of the DCT algorithm; different `libjpeg` settings have been used for pre-compression and unweighted JPEG WS estimation of $N = 1,600$ covers; lower MAEs indicate better performance; note the log scale

## 6.3 Summary and Outlook

In this chapter, we have presented several enhancements to the WS method for quantitative steganalysis of LSB replacement. More precisely, we have proposed a refined cover model for never-compressed covers, based on gentle assumptions on spatial correlation in natural images and empirical investigations of the distribution of better and less predictable areas therein. Further, an alternative cover model for steganalysis of spatial domain image representations generated from JPEG pre-compressed covers has been introduced. All improvements together integrate as a conditional cover model in the mixture cover model framework. The resulting detectors almost always match or beat the performance of all alternative quantitative detectors for LSB replacement; in the case of JPEG pre-compression our method outperforms existing quantitative detectors by up to one order of magnitude. This assertion is supported by empirical data from large-scale data sets with controlled (downsampling, pre-compression quality) and empirical (different sources) heterogeneity as well as dedicated robustness tests where deemed appropriate.

Despite these achievements, there remains always scope for future investigation. For enhanced WS on never-compressed covers, the most rewarding aspect to tackle in future research is probably finding an optimal weighting scheme that must be somehow more adaptive to a given image's predictability histogram than just the linear normalisation of all weights. Efforts in this direction should also envisage dealing with parity co-occurrence and related phenomena, such as flat or saturated areas, in a consistent and unified way. All current variants of WS address this issue with a mixture of weighting tweaks and bias correction, the joint outcome of which is mathematically intractable; and it is likely that one can do better. For JPEG WS, possible next steps include the search for a better predictability measure compatible with the adopted predictor and further investigation on the estimation of the JPEG pre-compression quality for non-standard quantisation matrices. For WS in general, one can conceive new refinements of the cover model to well-understood preprocessing chains (e.g., resampling factors, explicit CFA grids in colour images), thereby deepening the decision tree of the conditional cover model further. This might also include predictor improvements by incorporating side information, such as source-specific noise patterns extracted and estimated from other covers (and possibly stego objects) of the same source (see [162] for related methods in the field of digital image forensics). Last but not least, studying the relation between WS and so-called 'structural detectors' (in particular, the aggregation of deviations from individual cover assumptions) might shed light into the open question as to whether WS, which in fact exploits the parity structure by the term $x_i^{(p)} - \overline{x}_i^{(p)}$ in the estimation equation, is really something different, or nothing more than a mathematically better founded representative of the class of 'structural detectors'.

# Chapter 7
# Using Encoder Artefacts for Steganalysis of Compressed Audio Streams

The ISO/MPEG 1 Audio Layer-3 (MP3) audio compression algorithm [30, 113] is probably one of the most recognised and far-reaching developments in the area of digital media processing. The MP3 format enables compression rates of about 1/10 of the size of uncompressed digital audio while degrading the audible quality only marginally. Together with faster computers the moderate complexity of the compression algorithm, due to which software implementations of MP3 encoders/decoders with acceptable performance even on low budget home computers soon became available, the format simplified the interchange of music. As a result, MP3 gained worldwide popularity among its users and at the same time it threatened the music industry's conventional business models. The popularity of the format created demand for encoding tools and opened a market for a variety of programs for different needs. Today we count hundreds of MP3 encoder front ends based on several dozens of encoding engines, ranging from proof of concepts to targeted products either tuned for high speed, or optimised to costly and flexible tools for professionals.

## 7.1 MP3 Steganography and Steganalysis

Against this backdrop, the MP3 format became an interesting cover format for steganography. MP3 files are particularly suitable for storing and transmitting secret messages for at least three reasons:

1. The high popularity of the format is an advantage, because exchanging common and widely used types of data is less conspicuous to an observer. For instance, sharing an MP3 file over the Internet is a completely common task and doing so is, albeit not always legal, a plausible form of network communication.
2. Typical MP3 file sizes range between 2 and 4 MB. This is more than for other common formats (e.g., text documents or small-scale photographs

as e-mail attachments). As all practical embedding functions suffer from small capacity, larger file sizes simplify the transmission of medium-sized secret messages (e.g., fax messages or small digital photographs). The inconveniences of splitting up messages to distribute them over different cover objects can be almost avoided with MP3 covers.

3. The nature of the lossy MP3 compression makes it attractive for steganographic use. Lossy compression during encoding introduces indeterminacy (from the steganalyst's point of view) that can be exploited to carry hidden information securely. However, this indeterminacy must not be overestimated if the steganalyst has access to the original medium, for instance, by acquiring the original recording of commercial music (cf. cover–stego-attacks in Sect. 2.5.1).

Compared to the suitability of MP3 files for steganography, the number of known steganographic tools for this format is still quite limited. MP3Stego[1] [184] is based on the 8hz-mp3 encoder, a Dutch student programming project [1]. The embedding operation of MP3Stego hides message bits in the parity of block lengths. Although this procedure is limited to a very low capacity, it is (under certain conditions, see below) detectable [234]. The detector is based on the analysis of statistical properties, more precisely, the variance of block lengths in the MP3 stream. UnderMP3Cover [195], another steganographic tool, is based on the ISO reference sources for MP3 encoders. Its embedding function embeds message bits into the least significant bits of block amplification attributes, which binds the capacity to comparably low rates as for MP3Stego. Both its LSB approach and the fact that no key is used leave obvious vulnerabilities. Finally, Stego-Lame [222] pursues another approach and embeds secret messages in the time domain of uncompressed PCM audio data. The amount of information is very small and the embedding function adds redundancy by means of channel coding so that the hidden message is robust to subsequent lossy MP3 compression. This tool never went beyond an experimental stage. Meanwhile, the corresponding entry on SourceForge.net, an open source code sharing and development platform, has been removed due to inactivity. Aside from these publicly known stego tools we have to assume that some more are being used in practice. Although the complexity of MP3 compression exceeds those of typical steganographic tools (e.g., LSB replacement in images), the availability of commented source codes for MP3 encoders facilitates the composition of derivatives with steganographic extensions.

---

[1] As a convention, encoder names are printed in typewriter style.

## 7.1.1 Problem Statement in the Mixture Cover Model Framework

The situation for steganalysis of MP3 files is like a textbook example for the mixture cover model. The existing detector against MP3Stego can distinguish MP3 files with and without steganographic content quite reliably if the files are encoded with either MP3Stego or its underlying encoding engine 8hz-mp3. However, files from other encoders tend to have similar statistical properties to stego objects from MP3Stego, and thus are identified as false positives. Hence, the reliability of the detection algorithm largely depends on prior knowledge about the encoder of a particular file. While this situation might be sufficient for an academic detector or proof of concept, it is definitely not optimal for real-world applications. In practice, we usually cannot expect any prior knowledge about the source of arbitrary MP3 files.

To put it in the words of the mixture cover model: the actual encoder used to create an MP3 file is an unobservable random variable $H$ and the steganalyst can improve the detection reliability by estimating the realisation of $H$ before running the actual detector. Therefore, a procedure is required to determine the encoder of MP3 files based on characteristics that are typical for a certain implementation of the MP3 format specification. Ideally, the encoder classification shall be independent of the dimensions analysed by the actual detection method (cf. Eq. (3.25)). In the remainder of this chapter we present such a classification method and demonstrate its usefulness for more reliable steganalysis.

## 7.1.2 Level of Analysis and Related Work

The proposed method employs a naïve Bayes classifier (NBC) in conjunction with deliberately designed statistical features of compressed MP3 streams. So, the development of such a method largely depends on the discovery of suitable features. To find some, we systematically revisited the specification of MP3 compression (cf. Sect. 2.6.4.2 for a brief overview) and the architecture of typical encoder implementations with regard to degrees of freedom in the ISO specification that leave space for different interpretations by different implementations. It is the ambiguity in the standard that leads to different output streams for the same input data.

In order to perform a statistical characterisation of MP3 encoders, we have to find differences in the encoding process. These differences may originate from multiple sources. At the very first glance, all vaguely defined parameters in the specification are subject to different interpretations. However, the ISO/IEC standard precisely describes a large set of critical parameters, including the exact coefficients for the filter bank and threshold values for the psycho-acoustic model. Nevertheless, some implementations seem to vary or fine-tune these parameters. In addition, performance considerations may have led to sloppy implementations of the standard, such as through shortcuts in the inner quantisation loop or the choice of non-optimal Huffman tables. Furthermore, a number of parameter defaults for metainformation are up to the implementation (e.g., the *serial copy management system* (SCMS) flags, also known as protection bits [109]). The combinations of all these variations cause particular, and to a large extent identifying, features in the output stream that form indications for a specific encoder and therefore are subject to a detailed analysis.

Table 7.1: Structure of MP3 encoding and possibilities to extract features

| Transformation layer | Modelling layer | Bitstream layer |
|---|---|---|
| **Functionality** | | |
| filter bank | quantisation loop | add auxiliary data |
| MDCT transform | model decisions | set frame header bits |
| FFT transform | table selection | checksum calculation |
| | | stream formatting |
| **Points for analysis** | | |
| frequency range | size control | surface information |
| filter noise | model decisions | SCMS protection bit |
| audible artefacts | capability usage | SCMS original bit |

To structure the domains in which implementation-specific particularities in the MP3 encoding process can be expected, we subdivide the process into three layers as shown in Table 7.1. The *transformation layer* includes all operations that directly affect the audio data, namely the filter bank, and the MDCT and FFT time to frequency transformations, respectively. In this layer, variations in the filter coefficients or in the precision of the floating point operations may generate measurable features, such as typical frequency ranges or additional noise components.

We define all lateral components of the compression algorithm as part of the *modelling layer*. These subprocesses are less close to the underlying audio data and mainly perform the trade-off between size and quality of the compressed data. In this layer, encoder differences essentially occur in three ways:

1. Calculation of size control quantities, e.g., whether net or gross file sizes are used as reference for the bit rate control.

2. Model decisions: different threshold values lead to different marginal distributions of control parameters over the data stream.
3. Capability usage: some encoders do not support all compression modes specified in the MP3 standard.

The third layer, which we call *bitstream layer*, handles the formatting of already-compressed MP3 blocks into a valid MPEG stream. These operations include the composition of frame headers, the optional calculation of CRC checksums for error detection, and the insertion of metadata. For instance, quasi-standardised ID3 tags [178] contain information about the names of artists, interpreters, and publishers of audio files. Optional *variable bit rate* (VBR) headers store additional data which some MP3 players evaluate to display valid progress bars and enable efficient skipping within MP3 files with variable bit rate.[2] The existence of a certain kind of metainformation and its default values may be used as an indicator for the encoding program.

EncSpot [53], the only tool for MP3 encoder detection we are aware of, relies on the deterministic surface parameters of the bitstream layer. As these parameters are easily accessible, it is also simple to erase or change their values in order to deceive this kind of encoder detection. Therefore, we decided to use statistical features related to deeper structures of the encoder, because these features are much more difficult to manipulate. Our initial experiments with parameters of the transformation layer showed that those tend to depend largely on the type of audio data used as input. For example, it is impossible to measure encoder characteristics, such as the upper end of the frequency range, if the encoded audio material does not use the full range. Also, artefacts occur at typical envelopes or frequency changes that do not appear similarly in all kinds of music or speech. Hence, we decided to focus our level of analysis on the modelling layer, which promises to deliver the most robust features in terms of input signal independence and difficulty of manipulation.

## 7.1.3 Method

To precisely describe the nature of the features, we introduce additional formal notations. We denote a medium $x$ as $\dot{x}$ for the source (i.e., uncompressed) representation and as $x_i = \mathsf{Encode}_i(\dot{x})$ if it is encoded with encoding program $\mathsf{Encode}_i$, $1 \leq i \leq N$. We write the set of all files encoded with $\mathsf{Encode}_i$ as

---

[2] As MP3 has been specified for *constant bit rates* (CBRs), the majority of MP3 files are encoded as CBR with one of the predefined rates. However, some encoding programs optionally encode each frame with a different bit rate (out of the predefined rates), thus allowing *variable bit rate* (VBR) streams with baseline MP3.

$\mathcal{X}_i = \left\{ \boldsymbol{x} \mid \boldsymbol{x} = \mathsf{Encode}_i(\dot{\boldsymbol{x}}) \wedge \dot{\boldsymbol{x}} \in \dot{\Omega} \right\}$, where $\dot{\Omega}$ is the set of all uncompressed source media.

Function $\mathsf{Fx}(\boldsymbol{x})$ extracts a discrete or continuous feature $f$ from $\boldsymbol{x}$. The vector $\boldsymbol{f}$ of $k$ different features

$$\boldsymbol{f} = \mathsf{Features}(\boldsymbol{x}) = (\mathsf{Fx}_1(\boldsymbol{x}), \dots, \mathsf{Fx}_k(\boldsymbol{x})) \tag{7.1}$$

is called 'feature vector'. The elements constituting the feature vector[3] $\boldsymbol{f} \in \mathcal{F}^k$ are composed to be as similar as possible for different media $\boldsymbol{x} \in \mathcal{X}_i$ encoded with the same encoder $\mathsf{Encode}_i$, and at the same time as dissimilar as possible from all tuples of encoded media $(\boldsymbol{x}_i, \boldsymbol{x}_j) \in \mathcal{X}_i \times \bigcup_{j \neq i} \mathcal{X}_j$ encoded with different encoders. This way, the information on the characteristics of the encoder is consolidated in the value of $\boldsymbol{f}$.

Classifiers are algorithms which automatically classify an object, i.e., assign it to one of several predefined classes, according to its features. From the many approaches that can be found in the literature (cf. Sect. 2.9.2), we chose a classifier based on Bayesian logic as it seemed most suitable to handle mixed vectors of discrete and continuous features [156]. We show in Sect. 7.3 that notably accurate results are achievable with the simple *naïve Bayes classifier* (NBC) [49].

We use a classifier $\mathsf{Classify} : \mathcal{F}^k \to \mathbb{Z}$ to establish the relation between a specific realisation of $\boldsymbol{f} = \mathsf{Features}(\boldsymbol{x}_i)$ and the encoding program $\mathsf{Encode}_i$ used to create $\boldsymbol{x}_i$. If we do not have any knowledge of the encoder, we can only derive probabilistic evidence for this assignment. For a given medium $\boldsymbol{x}$, a classifier tries to compute the conditional probabilities

$$\mathsf{Prob}\left(\mathsf{Encode}_i | \mathsf{Features}(\boldsymbol{x})\right) = \mathsf{Prob}\left(\mathsf{Encode}_i | f_1 = \mathsf{Fx}_1(\boldsymbol{x}), \dots, f_k = \mathsf{Fx}_k(\boldsymbol{x})\right), \tag{7.2}$$

with $1 \leq i \leq N$, and then selects the most likely encoder $\mathsf{Encode}_i$, so that

$$\mathsf{Classify}\left(\mathsf{Features}(\boldsymbol{x})\right) = \arg\max_i \mathsf{Prob}(\mathsf{Encode}_i | \mathsf{Features}(\boldsymbol{x})). \tag{7.3}$$

The classifier's performance depends on its parameterisation, which can be induced from data. Therefore we assemble a training set $\mathcal{T}$ of tuples $t \in \mathbb{Z} \times \mathcal{F}^k$

$$\mathcal{T} = \left\{ (i, \mathsf{Encode}_i(\dot{\boldsymbol{x}})) \mid 1 \leq i \leq N \wedge \dot{\boldsymbol{x}} \in \dot{\Omega}_{\mathrm{sample}} \subset \dot{\Omega} \right\}. \tag{7.4}$$

Each element in $\mathcal{T}$ contains a consolidated representation of medium $\boldsymbol{x}_i$ and a reference to the known encoding program. We write a classifier trained with the training set $\mathcal{T}$ as $\mathsf{Classify}_\mathcal{T}$. The encoder predictions of a specific feature vector $\boldsymbol{f}$ and of an underlying medium $\boldsymbol{x}$ are denoted as $\mathsf{Classify}_\mathcal{T}(\boldsymbol{f})$ and $\mathsf{Classify}_\mathcal{T}(\mathsf{Features}(\boldsymbol{x}))$, respectively. To evaluate the quality

---

[3] We slightly abuse the power notation for the feature space. In fact, the domain $\mathcal{F}$ of $\mathsf{Fx}$ is not necessarily the same for all dimensions. This is so, for example, when discrete and continuous features are mixed.

of the classification, we regard the proportion $q$ of correctly classified cases[4] when the classifier is applied to elements of a test set $\mathcal{S}$, which is composed similarly of the training set $\mathcal{T}$:

$$q_{\text{Classify}}^{\mathcal{S}} = \frac{\left|\{(i, \boldsymbol{x}_i) \in \mathcal{S} \mid i = \text{Classify}\left(\text{Features}(\boldsymbol{x}_i)\right)\}\right|}{|\mathcal{S}|}. \tag{7.5}$$

As a weak form of reliability evaluation, the same training set $\mathcal{T}$ can be reclassified (within-sample performance), i.e., $\text{Classify}_{\mathcal{T}}(\text{Features}(\boldsymbol{x}_i))$ with $(i, \boldsymbol{x}_i) \in \mathcal{T}$. A stronger, more critical and more generalisable measure can be obtained from disjoint training and test sets, so that $\mathcal{S} \cap \mathcal{T} = \varnothing$ (out-of-sample performance).

NBCs make relatively strong assumptions, namely that all features are mutually independent. This means that the probability of the class conditional on the realisation of one feature does not depend on the realisation of any other feature:

$$\text{Prob}\left(\text{Encode}_i | \text{Fx}_j\right) = \text{Prob}\left(\text{Encode}_i | \text{Fx}_j \wedge \text{Fx}_\iota\right)$$
$$\forall (i, j, \iota) \in \{(i, j, \iota) | 1 \leq i \leq N, 1 \leq j, \iota \leq k \wedge j \neq \iota\}. \tag{7.6}$$

The advantage of this strong (and in practice rarely met) assumption is a small parameter space and a simple training and classification scheme: calculating the posterior probabilities is nothing more than computing weighted sums of attribute loadings. As a matter of fact, in many practical applications, the method has turned out to be quite robust even against violations of the independence assumption. A comprehensive evaluation of different classifiers by Langley et al. [149] concludes that the simple NBC performed equally or better than more complex classification methods for many realistic decision problems. Our own experiments with alternative classifiers are coherent with this result. We refrain from repeating details of the NBC training and classification procedure, as it is a widely known standard tool. Interested readers are referred to the literature, e.g., [49].

## 7.2 Description of Features

As a result of iterative comparisons and analyses of MP3 encoder differences, we discovered a number of features, of which we selected ten for encoder classification. For a structured presentation, the features are assigned to categories which are discussed separately in the following subsections.

---

[4] Unlike in chapters dealing with JPEG steganography, symbol $q$ does not denote compression quality.

## 7.2.1 Features Based on the Compression Size Control Mechanism

Distinct encoders seem to differ in the way the target bit rate is calculated, as we discovered measurable differences in the effective bit rate. According to the MP3 standard, each block can be encoded with one of 14 predefined bit rates between 32 and 320 kbps. However, because of the difficulty of reaching the target compressed size exactly, these rates act merely as guiding numbers. Some encoders treat them as upper limits, others as average. Also, the encoders differ in which fields of the frames are included in the length calculation performed in the compression loop. If the size of adjacent frames is included, or fixed headers at the beginning of MP3 files are counted as well, then the effective bit rate varies with the overall file size and converges to a target value with increasing number of frames. For example, the effective bit rates $\omega_{\text{eff}}$ of both `8hz-mp3` and `mp3comp` depend on the number of frames $\#_{\text{fr}}$, while there is no influence for files encoded with `lame` or `fhgprod`. We calculate the effective bit rate as

$$\omega_{\text{eff}} = \frac{(\text{length}_{\text{file}} - \#_{\text{junkbytes}} - \text{length}_{\text{meta information}}) \cdot 8\,\text{bits} \cdot 44.1\,\text{kHz}}{1152 \cdot \#_{\text{fr}}}.$$
(7.7)

The constants 44.1 kHz and $1{,}152$ are the sampling frequency and the frame length (in samples, see Sect. 2.6.4.2), respectively. Even for large files, we observe a measurable difference in the marginal $\omega_{\text{eff}}$ between all four encoders. To derive a bit rate feature based on this observation, we calculate a criterion $\kappa_1$ as the ratio between the effective bit rate $\omega_{\text{eff}}$ and the nominal bit rate $\omega_{\text{nom}}$:

$$\kappa_1 = \frac{\omega_{\text{eff}}}{\omega_{\text{nom}}}, \quad \text{with} \quad \omega_{\text{nom}} = \frac{1}{\#_{\text{fr}}} \sum_{i=1}^{\#_{\text{fr}}} \omega_{\text{nom}}^{(i)}, \tag{7.8}$$

where $\omega_{\text{nom}}^{(i)}$ is the nominal bit rate given in the header of the $i$th frame. To map this ratio to a symbolic feature $f_1$, we define the extraction function $\mathsf{Fx}_1$ as follows:

$$\mathsf{Fx}_1(x) = \begin{cases} 1 & \text{for} & \kappa_1 < 1 - 1 \cdot 10^{-4} \\ 2 & \text{for } 1 - 1 \cdot 10^{-4} \leq \kappa_1 \leq 1 \\ 3 & \text{for} & 1 < \kappa_1 \leq 1 + 5 \cdot 10^{-6} \\ 4 & \text{otherwise.} \end{cases} \tag{7.9}$$

The number of levels and the exact boundaries for this feature, as well as for the following ones, are determined by an iterative process of comparing a set of test audio files. We report the functions which led to the best experimental results, even though we acknowledge that many decisions are a bit arbitrary and further fine-tuning may still be possible.

In Sect. 2.6.4.2, we have mentioned that an MP3 stream consists of a sequence of frames and that two variable-size granules constitute a frame of fixed size. The quantisation loop adjusts the size of the granules separately according to two criteria:

1. Size: the granule must fit into the available space.
2. Quality: compression noise shall remain imperceptible.

For some encoders, e.g., **shine**, we observed a slight bias for quality over size. As the 'hard' space limit counts for both granules together, the first granules $g_1^{(i)}$ of all frames ($1 \leq i \leq \#\text{fr}$) tend to grow larger than the second ones $g_2^{(i)}$. Hence, we measure the proportion of frames in the file where the size of the first granule $|g_1|$ exceeds the size of the second granule $|g_2|$:

$$\kappa_2 = \frac{1}{\#\text{fr}} \sum_{i=1}^{\#\text{fr}} \delta_{+1,\text{sign}(|g_1^{(i)}|-|g_2^{(i)}|)}. \tag{7.10}$$

Note that the granule bias function measure needs to be modified slightly for stereo files to ensure that the blocks of the left and right channels are compared separately. Again, we define a mapping function, now for feature $f_2$:

$$\text{Fx}_2(x) = \begin{cases} 1 & \text{for} & \kappa_2 < 0.50 \\ 2 & \text{for} & 0.50 \leq \kappa_2 < 0.55 \\ 3 & \text{for} & 0.55 \leq \kappa_2 < 0.70 \\ 4 & \text{otherwise.} \end{cases} \tag{7.11}$$

The next feature makes use of characteristics of the reservoir mechanism. We found that the acceleration of the rise in reservoir usage between silent and dynamic parts in the audio stream differs between some encoders. Yet other encoders do not even use the reservoir. As the vast majority of audio files start with a tiny silence, we derive the feature $f_3$ from the number of reservoir bytes shared between the first and second frames, $\text{res}_{(1,2)}$:

$$\text{Fx}_3(x) = \begin{cases} 1 & \text{for} & \text{res}_{(i,i+1)} = 0 & \forall i: 1 \leq i < \#\text{fr} \\ 2 & \text{for} & \text{res}_{(1,2)} > 300 \\ 3 & \text{otherwise.} \end{cases} \tag{7.12}$$

Function $\text{Fx}_3$ returns a value of 1 if the reservoir is not used in the entire file. Values two and three indicate hard and soft reservoir usage, respectively.

The last feature in this category is less justified theoretically, but our evaluation shows that it has some impact on a better separation between two versions of the Xing encoder, namely **xing98** and **xing3**. We observed that **xing3** uses a different size control mechanism for the second block of every granule of stereo files. According to the ISO/MPEG 1 Audio Layer-3 terminology [113], *big values* are spectral coefficients with absolute values after quantisation greater than 1. The average number of big value coefficients is a valid indicator for the extent of size reduction in the quantisation loop.

To derive a continuous feature from the different spread of histogram values in the stereo channels, we measure the entropy from the histogram with the approximation given in [170]:

$$H \approx -\sum_{j=1}^{d_{\max}} d_j \log_2 d_j + \log_2 \Delta, \tag{7.13}$$

with $d_j$ denoting the number of occurrences in the $j$th bin and $\Delta$ as bin size. Since $\Delta$ is constant for all encoders, we use a simplified function to calculate feature $f_4$:

$$\mathsf{Fx}_4(\boldsymbol{x}) = -\sum_{j=1}^{60} d_j \log_2 d_j. \tag{7.14}$$

Note that in contrast to previous features, $\mathsf{Fx}_4$ is a continuous feature that is modelled by the classifier as a Gaussian random variable with mean $\mu_i$ and standard deviation $\sigma_i$ for the $i$th encoder $\mathsf{Encode}_i$. However, as this feature evaluates the characteristics of the second channel in stereo data, it is not applicable to mono files; hence, we cannot discriminate between xing3 and xing98 for mono files.

### 7.2.2 Features Based on Model Decisions

The psycho-acoustic model is a second source for distinguishing features. Differences in the computation of control parameters or modifications in the choice of threshold values lead to encoder-specific marginal distributions of field values over all frames, granules or blocks in a file.

The binary pre-emphasis flag controls an optional additional amplification of high frequencies and is individually set for each compressed block $\boldsymbol{b}_i{}^5$ ($1 \leq i \leq \#\mathrm{bl}$, with $\#\mathrm{bl}$ as the number of blocks in a file). The ISO/MPEG 1 Audio Layer-3 standard explicitly leaves latitude on when to set this flag:

The condition to switch on the preemphasis is up to the implementation. [113, p. 110]

As a result, different encoders treat this flag differently. This makes it easy to derive an operable feature by calculating the proportion of blocks with pre-emphasis flag set:

$$\kappa_5 = \frac{1}{\#\mathrm{bl}} \sum_{i=1}^{\#\mathrm{bl}} \mathsf{Preflag}(\boldsymbol{b}_i). \tag{7.15}$$

---

[5] We reuse symbol $\boldsymbol{b} \in \mathcal{B}$ for MP3 blocks in this chapter. It should not be confused with bit vectors in syndrome coding as introduced in Sect. 2.8.2, or with filter coefficients of Sect. 6.1.1, or with the binomial distribution function.

Function Preflag : $\mathcal{B} \to \{0, 1\}$ extracts the pre-emphasis flag of a given block. Criterion $\kappa_5$ can be mapped to the symbolic feature $f_5$ as follows:

$$\mathsf{Fx}_5(x) = \begin{cases} 1 & \text{for} & \kappa_5 = 0.00 \\ 2 & \text{for} & 0.00 < \kappa_5 \leq 0.01 \\ 3 & \text{for} & 0.01 < \kappa_5 \leq 0.05 \\ 4 & \text{for} & 0.05 < \kappa_5 \leq 0.10 \\ 5 & \text{for} & 0.10 < \kappa_5 \leq 0.21 \\ 6 & \text{for} & 0.21 < \kappa_5 \leq 0.35 \\ 7 & \text{for} & 0.35 < \kappa_5 \leq 0.62 \\ 8 & \text{for} & 0.62 < \kappa_5 \leq 0.77 \\ 9 & \text{otherwise.} \end{cases} \tag{7.16}$$

Our experiments suggest that the symbolic interpretation of $f_5$ leads to better classification results than a treatment as continuous feature with assumed Gaussian distribution. Also, the bounds for the symbol assignment have been determined experimentally.

The MP3 audio format supports different block types which enable an optimal trade-off for audio sequences that demand a higher time resolution at the cost of frequency resolution, and vice versa. In a typical MP3 file, the majority of blocks are encoded with block type 0, the *long block* with low time and high frequency resolution. Block type 2 defines a *short block*, which stores fewer coefficients for three different points in time. Two more block types are specified to perform smooth shifts between the two extreme types 0 and 2. Hence, the standard defines a graph of valid block transitions between two subsequent blocks $b_i$ and $b_{i+1}$, as illustrated in Figure 7.1.

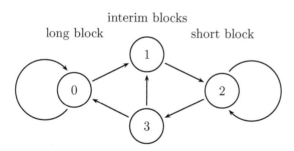

Fig. 7.1: Directed graph of valid MP3 block type transitions

An evaluation of block type transitions of MP3 files from different encoders uncovers two interesting details. First, a number of encoders (shine, all xing*) apparently do not use short blocks at all and thus always encode with block type 0. Second, other encoders (lame, gogo, and plugger) allow 'illegal' transitions, mainly at the beginning of a file. As these transitions are rarely observable from other encoders, they can serve as reliable features for encoder identification. Using the notation $\breve{b}_i$ for the block type of the $i$th block in an MP3 file, we construct the extraction function for feature $f_6$ as follows:[6]

$$
\mathsf{Fx}_6(\boldsymbol{x}) = \begin{cases} 1 & \text{for} \quad \breve{b}_i = 0 \quad \forall\, i\colon 1 \le i \le \#_{\mathrm{bl}} & (\texttt{shine}, \texttt{xing}) \\ 2 & \text{for} \quad \breve{b}_1 = 0 \,\wedge\, \breve{b}_2 = 2 & (\texttt{lame}) \\ 3 & \text{for} \quad \breve{b}_1 = 2 \,\wedge\, \breve{b}_2 = 3 & (\texttt{gogo}) \\ \text{else} \begin{cases} 4 & \text{for} \quad |\{\boldsymbol{b}_i|\breve{b}_i = 2\}| = \\ & \qquad |\{\boldsymbol{b}_i|\breve{b}_i = 3\}| = 1 & (\texttt{plugger}) \\ 5 & \text{otherwise.} & (\text{all other encoders}) \end{cases} \end{cases}
\tag{7.17}
$$

Since we cannot explain these strange transitions, we conjecture that they are either bugs or intentionally placed to leave an encoder trace in the output data.

## 7.2.3 Features Based on Capability Usage

The third category of features exploits the fact that some encoders do not implement all functions specified in the MP3 standard. We call this category *capability usage* and clearly distinguish these capabilities from what we call 'surface parameters', such as header flags, because the latter can easily be changed without touching the compressed data.

The *scale factor selection information* (SCFSI) is a parameter that allows an encoder to reuse scale factors of the first granule for the second granule of a block. However, only few encoders make use of the possibility to share scale factors in a frequency-band specific manner, namely `lame`, `gogo`, and `xingac21` ('AudioCatalyst'). To define feature $f_7$, reflecting the elaborate use of SCFSI, let $\mathcal{Q}_{\mathrm{SCFSI}}$ be the set of all possible combinations of SCFSI flags and $\mathcal{Q}'_{\mathrm{SCFSI}} \subset \mathcal{Q}_{\mathrm{SCFSI}}$ the set of combinations used in a particular file:

$$
\mathsf{Fx}_7(\boldsymbol{x}) = \begin{cases} 1 & \text{for} \quad |\mathcal{Q}'_{\mathrm{SCFSI}}| \le 2 \\ 2 & \text{otherwise.} \end{cases}
\tag{7.18}
$$

MP3 frames have a fixed size, but it serves as an upper limit in the bitstream specification only. The actual amount of information used to describe a specific audio signal may vary. We refer to this quantity as *effective frame size* $s_{\mathrm{eff},i}^{\mathrm{fr}}$. The MP3 standard imposes no constraints on the effective frame size to match a multiple of bytes, words or quad-words. However, we observed that some encoders (`8hz-mp3`, `bladeenc`, `m3ec`, `plugger`, `shine`, `soloh`) adjust all effective frame sizes to byte boundaries, while others do not. We use this characteristic as feature $f_8$:

---

[6] We specify the function for mono files. Stereo files work similarly if blocks are evaluated in pairs.

$$\mathsf{Fx_8}(\boldsymbol{x}) = \begin{cases} 1 & \text{for} \quad s_{\text{eff},i}^{\text{fr}} \equiv 0 \ (\mathrm{mod}\ 8) \quad \forall\, i: 1 \le i \le \#_{\text{fr}} \\ 2 & \text{otherwise.} \end{cases} \tag{7.19}$$

The quantised MDCT coefficients are further compressed by a Huffman-style entropy encoder. In contrast to the general method proposed by Huffman [107], the tables are not computed in real time from the marginal symbol distribution. To avoid the transmission of marginal distributions or table data, the developers of MP3 standardised a set of 28 predefined Huffman tables that were empirically optimised for the most likely cases in audio compression. In rare cases of longer code words, an escape mechanism allows the encoder to store individual input symbols directly. An MP3 encoder chooses the most suitable table separately and independently for each of three so-called 'regions' of the big value MDCT coefficients. Although the optimal table selection can be found quite efficiently using marginal symbol distributions, some encoders apparently increase performance by using heuristics to quickly select a suitable table rather than the optimal one. From a comparison of table usage frequencies, we found two noteworthy characteristics. First, all Xing encoders seem to avoid using table number 0 for region 2 strictly.[7] Second, only a few encoders (m3ec, mp3enc31, uzura) use table 23 for the regions 1 and 2. We exploit these observations as additional information for our classification:

$$\mathsf{Fx_9}(\boldsymbol{x}) = \begin{cases} 1 & \text{for} \quad \sum_{i=1}^{\#\text{bl}} \delta_{0,\mathsf{HuffTable}(\boldsymbol{b}_i,2)} = 0 \\ 2 & \text{for} \quad \sum_{i=1}^{\#\text{bl}} \sum_{j=1}^{2} \delta_{23,\mathsf{HuffTable}(\boldsymbol{b}_i,j)} > 0 \\ 3 & \text{otherwise.} \end{cases} \tag{7.20}$$

Function $\mathsf{HuffTable} : \mathcal{B} \times \mathbb{Z} \to \mathbb{Z}$ extracts the Huffman table from a given block for a selected region. Also, shine uses only a subset of the defined tables. However, we refrain from adjusting this feature for the detection of shine, as we can already identify this rarely used encoder with several other features.

### 7.2.4 Feature Based on Stream Formatting

Our last feature is based on particularities of stream formatting and most similar to what we call 'surface features'. This means that it can be manipulated relatively easily without digging into the details of re-compression. Nevertheless, it is not as superficial as header bits.

Independently of whether the reservoir mechanism is used or not, there may be a couple of bytes unused and filled up to meet the fixed frame length. These so-called 'stuffing bits' can be set to arbitrary values. For a closer

---

[7] Following the conventions in the MP3 standard, we count the regions from 0.

examination of these values, we composed histograms of the byte values in the stuffing areas. While most encoders set all stuffing bits to 0, we still found some exceptions and mapped them into a symbolic feature $f_{10}$:

$$Fx_{10}(x) = \begin{cases} 1 & \text{for stuffing with 0s} \\ 2 & \text{for no stuffing at all (i.e., } s^{\text{fr}}_{\text{eff},i} \text{ always matches fixed size)} \\ 3 & \text{for stuffing with 0x55 or 0xaa} \\ 4 & \text{for stuffing with ``GOGO'' (alternating values 0x47 and 0x4f)} \\ 5 & \text{otherwise.} \end{cases}$$

(7.21)

We conclude this section with a general remark: the set of features presented in this section has been selected from a larger set of encoder-specific particularities in MP3 files which we discovered in the course of our analysis. Therefore we do not claim that this feature set is comprehensive, or 'optimal' in any sense. The experiments in the next section demonstrate that this feature set fits its intended purpose. Nevertheless, it is still feasible to find further differentiating features. Such features may be necessary to reliably separate new encoders, or encoders that were not included in our analysis.

## 7.3 Experimental Results for Encoder Detection

For our experimental work, we extended the $R$ language [110, 199] by a new package for statistical analyses of MP3 files based on the open source MP3 player mpg123 [103]. All results are based on an MP3 database of about 2,400 files encoded with 20 different encoders (see Table G.11 in Appendix G). The audio files were selected from different sources to make the measurements independent of specific types of music or speech. We included tracks from a remastered CD of 1998 Grammy nominees, from a compilation of *The Blues Brothers* movie soundtrack (including some live recordings), and from piano music by Chopin, as well as *sound quality assessment material* (SQAM) files with speech and instrumental sounds. All source files were imported from CD recordings and stored as PCM wave files with 44.1 kHz, 16 bit, stereo.

### 7.3.1 Single-Compressed Audio Files

If supported by the encoder, we converted every source audio file to MP3 with three constant bit rates that we believe are the most widely used rates for MP3 files (112, 128, and 192 kbps). A number of additional MP3 files with variable bit rates—with two quality settings each—were generated for the encoders that support VBR, namely iTunes, lame, and xingac21.

Table 7.2: Classifier performance on disjoint training and test data

**True encoder**

| Percent of files classified as ... | 8hz-mp3 | bladeenc | fastencc | fhgprod | gogo | iTunes | l3enc272 | l3encdos | lame | m3ec | mp3comp | mp3enc31 | plugger | shine | soloh | soundjam | uzura | xing3 | xing98 | xingac21 |
|---|---|---|---|---|---|---|---|---|---|---|---|---|---|---|---|---|---|---|---|---|
| 8hz-mp3 | 95 | | | | | | | | | | | | | | 23 | | | | | |
| bladeenc | | 100 | | | | | | | | | | | | | | | | | | |
| fastencc | | | 100 | | | | | | | | | | | | | | | | | |
| fhgprod | | | | 94 | | | | | | | 38 | | | | | | | | | |
| gogo | | | | | 100 | | | | | | | | | | | | | | | |
| iTunes | | | | | | 100 | | | | | 2 | | | | | | | | | |
| l3enc272 | | | | | | | 84 | | | | | | | | | | | | | |
| l3encdos | | | | | | | 16 | 100 | | | | | | | | | | | | |
| lame | | | | | | | | | 100 | | | | | | | | | | | |
| m3ec | | | | | | | | | | 100 | | | | | 3 | | | | | |
| mp3comp | | | | 6 | | | | | | | 62 | | | | | | | | | |
| mp3enc31 | | | | | | | | | | | | 95 | | | | | | | | |
| plugger | | | | | | | | | | | | | 100 | | | | | | | |
| shine | | | | | | | | | | | | | | 100 | | | | | | |
| soloh | 5 | | | | | | | | | | | | | | 74 | | | | | |
| soundjam | | | | | | | | | | | | | | | | 100 | | | | |
| uzura | | | | | | | | | | | | | | | | | 100 | | | |
| xing3 | | | | | | | | | | | | 3 | | | | | | 85 | 13 | |
| xing98 | | | | | | | | | | | | | | | | | | 15 | 87 | |
| xingac21 | | | | | | | | | | | | | | | | | | | | 100 |
| | 100 | 100 | 100 | 100 | 100 | 100 | 100 | 100 | 100 | 100 | 100 | 100 | 100 | 100 | 100 | 100 | 100 | 100 | 100 | 100 |

$|\mathcal{T}_2| = |\mathcal{T}_1 \setminus \mathcal{T}_2| \approx 1,200$ stereo files, $N = 20$ encoders, $k = 10$ features, total err. rate: 5.1%

To measure the performance of our proposed method, we implemented a naïve Bayes classifier (NBC) [49] for fixed feature vectors of both symbolic and continuous features. In the first experiment, we trained the classifier $\mathsf{Classify}_{\mathcal{T}_1}$ with a training set $\mathcal{T}_1$ of about 2,400 cases. For each case, we extracted a feature vector $\boldsymbol{f} = \mathsf{Features}(\boldsymbol{x}_i)$ from a file encoded with a defined encoder $\mathsf{Encode}_i$ and used these tuples to induce classification parameters for $\mathsf{Classify}_{\mathcal{T}_1}$. To evaluate the performance of $\mathsf{Classify}_{\mathcal{T}_1}$, we use the same feature vectors as input to the classifier and compare the predicted encoders to the known true values. In this experiment we achieve a success rate

Table 7.3: Classifier performance measured with hundredfold cross-validation

| | True encoder | | | | | | | | | | | | | | | | | | | |
|---|---|---|---|---|---|---|---|---|---|---|---|---|---|---|---|---|---|---|---|---|
| | 8hz-mp3 | bladeenc | fastencc | fhgprod | gogo | iTunes | 13enc272 | 13encdos | lame | m3ec | mp3comp | mp3enc31 | plugger | shine | soloh | soundjam | uzura | xing3 | xing98 | xingac21 |
| **Percent of files classified as …** | | | | | | | | | | | | | | | | | | | | |
| 8hz-mp3 | 95 | 0 | · | · | · | · | · | · | · | 0 | · | · | · | · | 20 | · | · | · | · | · |
| bladeenc | 0 | 100 | · | · | · | 0 | · | · | · | · | · | · | · | · | 0 | · | · | · | · | · |
| fastencc | · | · | 100 | · | · | 1 | · | · | · | · | · | 0 | · | · | · | 0 | · | · | · | · |
| fhgprod | · | · | · | 93 | · | · | 0 | 0 | · | · | 37 | · | · | · | · | · | · | · | · | · |
| gogo | · | · | · | · | 100 | · | · | · | · | · | · | · | · | · | · | · | · | · | · | · |
| iTunes | · | · | · | · | · | 99 | 0 | · | · | · | · | 2 | · | · | · | 2 | 0 | · | · | · |
| 13enc272 | · | 0 | · | · | · | · | 86 | 28 | · | · | · | 0 | · | · | · | · | · | · | · | · |
| 13encdos | · | · | · | · | · | · | 13 | 72 | · | · | · | · | · | · | · | · | · | · | · | · |
| lame | · | · | · | · | · | · | · | · | 100 | · | · | · | · | · | · | · | · | · | · | · |
| m3ec | 0 | · | · | · | · | · | · | · | · | 99 | · | · | 0 | · | 1 | · | · | · | · | · |
| mp3comp | · | · | · | 7 | · | · | 0 | · | · | · | 63 | · | · | · | · | · | · | · | · | · |
| mp3enc31 | · | · | 0 | · | · | · | · | · | · | · | · | 97 | · | · | · | 1 | · | 0 | 1 | · |
| plugger | · | · | · | · | · | · | · | · | · | 0 | · | · | 100 | · | · | · | · | · | · | · |
| shine | · | · | · | · | · | · | · | · | · | · | · | · | · | 100 | · | · | · | · | · | · |
| soloh | 5 | 0 | · | · | · | · | · | · | · | 0 | · | · | · | · | 79 | · | · | · | · | · |
| soundjam | · | · | · | · | · | · | · | · | · | · | · | 0 | · | · | · | 97 | · | · | · | · |
| uzura | · | · | · | · | · | 0 | · | · | · | · | · | 0 | · | · | · | · | 100 | · | · | · |
| xing3 | · | · | · | · | · | · | · | · | · | · | · | 0 | · | · | · | · | · | 88 | 11 | · |
| xing98 | · | · | · | 0 | · | · | 0 | · | · | · | · | 0 | 0 | · | · | · | · | · | 12 | 88 | · |
| xingac21 | · | · | · | · | · | · | · | · | · | 0 | · | 0 | · | · | · | · | · | · | · | 100 |
| | 100 | 100 | 100 | 100 | 100 | 100 | 100 | 100 | 100 | 100 | 100 | 100 | 100 | 100 | 100 | 100 | 100 | 100 | 100 | 100 |

$|T_{2_i}| = 1{,}000$ stereo files, $N = 20$ encoders, $k = 10$ features, total err. rate: 5.0%. Dots for no misclassifications, values of 0 indicate few occurrences rounded to 0 (i.e., $< 0.5\%$).

of $q^{T_1}_{\mathsf{Classify}_{T_1}} = 96.2\%$. As a measure of confidence, we also calculate the average posterior probability over the predicted encoders $\max_i \mathsf{Prob}\,(\mathsf{Encode}_i | f) = 96.1\%$.

To check the robustness of our results and to reduce the risk of tautological finding, we repeated the experiment with a split-half method. We trained the classifier $\mathsf{Classify}_{T_1}$ with a subset $T_2 \subset T_1$ of the first training set $T_1$. All other elements from $T_1 \setminus T_2$ constitute the test set. The results of this second evaluation are shown in Table 7.2. We found an overall hit rate of $q^{T_1 \setminus T_2}_{\mathsf{Classify}_{T_2}} = 94.9\%$ and an average posterior probability of 95.9%. As both quality measures differ only marginally from the first experiment ($-1.3$ and

−0.2 percentage points, respectively), we conclude that the proposed method can also reliably identify the encoders of unseen MP3 files. (The difference between 'unseen' and 'unknown' data is that the origin of the former is controlled by the researcher, but not fed into the classifier, whereas the source of the latter is uncertain even to the researcher.)

A closer look at the results shows that the main sources for classification errors occur between closely related encoding engines, such as the DOS and UNIX versions of Fraunhofer's l3enc, and between two subsequent versions of Xing encoders (xing3 and xing99). Also, soloh produces false classifications, mostly towards 8hz-mp3, and especially for source files from one CD with a comparatively low recording level.

To gain even more confidence in what has been reported in [25] and [26], we also conducted a hundredfold cross validation, where in each iteration a random set of $|\mathcal{T}_2| = 1,000$ has been used for training and all other feature vectors have been for testing. The results largely support our conclusions, with an average overall hit rate $1/100 \sum_{i=1}^{100} q_{\mathsf{Classify}_{\mathcal{T}_{2_i}}}^{\mathcal{T}_{1_i} \setminus \mathcal{T}_{2_i}} = 95.0\%$; that is even slightly above the figures we reported in previous publications. Detailed results are reported in Table 7.3.

## 7.3.2 Importance of Individual Features

Aside from the overall classification performance, it is also interesting to look at the contribution of individual features. Table 7.4 summarises the features proposed in Sect. 7.2. We use a jackknife method to evaluate the importance of each feature for the classification result empirically: training and classification is repeated several times, thereby excluding individual features one by one. The additional overall classification error is a measure of the importance of a feature. According to this measure, the effective bit rate seems to be the most important feature, followed by reservoir usage.

## 7.3.3 Influence of Double-Compression

This section deals with the special case when MP3 files are compressed and re-compressed several times with different encoders. This can happen, for instance, when an existing audio stream is edited or resampled at a different, usually lower, bit rate. To extend the notation from Sect. 7.1.3, we write

$$x_{i,j} = \mathsf{Encode}_j\left(\mathsf{Decode}(x_i)\right) = \mathsf{Encode}_j\left(\mathsf{Decode}\left(\mathsf{Encode}_i(\dot{x})\right)\right), \ 1 \leq i, j \leq n, \tag{7.22}$$

as a double-compressed medium with the encoding sequence $(i, j)$. Here, $\mathsf{Decode} : \Omega \to \dot{\Omega}$ is a decoding function that converts MP3 streams to

Table 7.4: Overview of features used for classification

| Feature Description | Levels | Impact [a] |
|---|---|---|
| **Size control features** | | |
| $Fx_1$  Effective bit rate ratio | 4 | 8.35 |
| $Fx_2$  Granule size balance | 4 | 0.08 |
| $Fx_3$  Reservoir usage ramp | 3 | 5.01 |
| $Fx_4$  Entropy of big MDCT coefficients *(continuous)* | | 2.15 |
| **Model decision features** | | |
| $Fx_5$  Pre-empahsis flag ratio | 9 | 1.73 |
| $Fx_6$  Block type transitions | 5 | 1.56 |
| **Capability usage features** | | |
| $Fx_7$  Scale factor selection information | 2 | 0.50 |
| $Fx_8$  Frame length alignment | 2 | 0.92 |
| $Fx_9$  Huffman table selection | 3 | 0.63 |
| **Stream formatting feature** | | |
| $Fx_{10}$  Stuffing byte values | 5 | 0.88 |

[a] Drop in classification performance if feature excluded (in percentage points). Higher values indicate a more important contribution to a correct classification.

uncompressed PCM audio data. The fact that our features are extracted from the modelling layer suggests that the last encoder might dominate the classification decision regardless of the previous processing of the underlying audio data. Accordingly, our hypothesis is $\mathsf{Classify}(\mathsf{Features}(\boldsymbol{x}_{i,j})) = j$. However, this is hard to support analytically and there may be some exceptions for pathologic signals. So, we pursue an experimental approach to answer this question. A sample of 250 randomly selected MP3 files from the test database including all 20 encoders with their respective bit rate variations has been decompressed with the decoder MPG123 [103]. Then the resulting PCM files were re-compressed with each of five arbitrarily chosen encoders, leading to a new set of 1,250 double-compressed MP3 files.[8] The bit rate has been kept constant at 128 kbps for the second compression. Table 7.5 reports the results from a classification of the double-compressed test files.

The figures support our hypothesis: the second encoder dominates in the overwhelming majority of cases. Still, there are differences between encoders $j$ that can be explained by the fact that the proposed classification method

---

[8] The selection of the encoders used for re-compression was based on practical aspects, mainly the effort required to automate the whole procedure as a batch job.

Table 7.5: Classification of double-compressed MP3 material

| | Medium $x_{i,j}$ classified as ... | | |
|---|---|---|---|
| Second encoder | first encoder $(i)$ | second encoder $(j)$ | other $\notin \{i,j\}$ |
| gogo | 0.0% | 100.0% | 0.0% |
| shine | 0.9% | 99.1% | 0.0% |
| lame | 0.9% | 98.2% | 0.9% |
| 8hz-mp3 | 0.0% | 93.6% | 6.5% |
| bladeenc | 4.6% | 88.0% | 7.9% |

Base: 250 MP3 files per row

can identify some encoders more reliably than others. Although we tried to find systematic relations with a further drill-down into the cases where the second encoder has not been identified correctly, no signs of dependence between source encoders or encoding parameters were discovered. Albeit unlikely, there may also be influences from the characteristics of decoder Decode applied for the creation of double-compressed data. Since a comprehensive evaluation of double-compression with statistical means requires a considerable sample size of about 30,000 test files, we refrained from going into more detail. We thus consider the dominance of the second encoder as a valid and operable rule of thumb.

# 7.4 Experimental Results for Improved Steganalysis

To demonstrate the advances in steganalysis due to pre-classification, we assembled a test set of 500 clean MP3 covers from different encoders together with 369 stego objects from MP3stego [184], a steganographic extension to the 8hz-mp3 encoder. The idea behind the detector of MP3stego steganography as described by Westfeld [234] is quite simple. Since the embedding function embeds message bits into the LSB of the block size by repeatedly re-iterating the compression loop until the LSB has the correct semantic, the otherwise unchanged reservoir mechanism generates a higher variance in block sizes than in typical 8hz-mp3 output files. This variance criterion can be exploited to build a detector.

If we run this detector against MP3stego directly on the test set, we clearly identify all 369 stego objects, but face an additional 377 false positives (75.4%). This is so because encoders other than 8hz-mp3 produce higher block size variance by default, supposedly in an attempt to adjust the size

control better to the local audio content. The proposed encoder classification method can be employed to filter all files from other encoders except
8hz-mp3. This pre-classification removed all false alarms while still 312 stego
objects were reliably pre-classified and detected. Bearing in mind that false
positives are worse than misses, this is already a considerable improvement.

The miss rate of 15% can be diminished further to zero by including the
*original bit*, an SMCS header flag part of the bitstream layer, into the classification decision. Even though this is a deviation from the strict criteria
postulated in Sect. 7.1.2, we believe that the prospect of perfect steganalysis
(i.e., no detection error at all) justifies the drawback that this flag is easier
to forge than other features. Summarising the results for steganalysis, we
conclude that only in combination with source classification do the detection methods have sufficient discriminative power for a large scale search for
stego objects in MP3 files. This constitutes another practical example for the
concepts behind the mixture cover model proposed in Chapter 3.

## 7.5 Explorative Analysis of Encoder Similarities

Possible applications of encoder classifications are not limited to more reliable steganalysis and multimedia forensics. In the light of our theory of
Chapter 3, a feature set can be interpreted as a cover model. Our MP3 feature set thus constitutes a model for MP3 encoder artefacts. By calculating
appropriate distance metrics, we are in a position to draw conclusions about
the (dis)similarity of encoder implementations.

From ad hoc research of Internet sources we know that the encoders included in this study are not independent developments, but rather different
branches of a few core implementations. Apart from these explicit links, we
gained further indications from an analysis of encoding programs. For example, triggered by the misclassifications of soloh and 8hz-mp3, we analysed
the binary of soloh and found references in text strings to an early version
of 8hz-mp3. The following experiment is a demonstration of the idea that
similarity in statistical features may reveal insights about the 'intellectual
origin' of certain encoders.

To quantify the similarities between different encoders, we can exploit the
parameters of a trained classifier. The parameter set of the NBC consists of
$k$ matrices $\boldsymbol{\theta}^{(l)}, 1 \leq l \leq k$, one per feature. For symbolic features, $\boldsymbol{\theta}^{(l)}$ stores
the conditional probabilities that feature $\mathsf{Fx}_l$ takes value $f$ if medium $\boldsymbol{x}$ has
been created with encoder $\mathsf{Encode}_i$,

$$\boldsymbol{\theta}_{i,f}^{(l)} = \mathsf{Prob}\left(\mathsf{Fx}_l(\boldsymbol{x}) = f | \boldsymbol{x} \in \mathcal{X}_i\right). \tag{7.23}$$

For continuous features, $\theta_{i,1}^{(l)} = \mu_i^{(l)}$ stores the mean and $\theta_{i,2}^{(l)} = \sigma_i^{(l)}$ the
standard deviation of the distribution of $f_l$ for encoder $\mathsf{Encode}_i$. We interpret

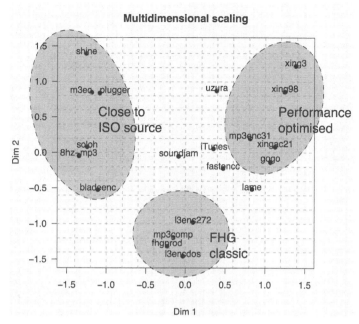

Fig. 7.2: Encoder similarity: Euclidean distance between parameters in $\boldsymbol{\theta}^{(l)}$

the elements of the collection of vectors $\boldsymbol{\theta}^{(l)}$ as points in Euclidean space and compute a distance matrix $\boldsymbol{o}$ as

$$o_{i,j} = \sqrt{\sum_{l=1}^{k} \sum_{f=1}^{\max(l|k)} \left(\theta_{i,f}^{(l)} - \theta_{j,f}^{(l)}\right)^2} \quad \text{with} \quad 1 \leq i, j \leq N. \tag{7.24}$$

The relation between encoders can be visualised as in Fig. 7.2 by employing a *multi-dimensional scaling* (MDS) procedure which projects $\boldsymbol{o}$ onto a two-dimensional space $\mathbb{R}^2$ while minimising the square error between $\boldsymbol{o}$ and the $L_2$-norm in the resulting plot [229]. Hence, the proximity of encoder locations in the figure indicates the similarity between statistical properties of encoder outputs with respect to our features.

Our prior knowledge of the close relation between `soloh` and `8hz-mp3` can be visually confirmed in Fig. 7.2. The respective points appear next to each other. Moreover, we identify that the encoders stick together in three clusters, as indicated by the grey bubbles. The encoders derived from the source code published in conjunction with the ISO standard are located in the upper left corner, whereas the 'official' encoders from the Fraunhofer Institute reside in the bottom area. The upper-right corner is populated with more performance-optimised consumer products. Lastly, the presumably more recent developments, such as `lame` and `iTunes`, appear in the middle area. Surprisingly, this result is coherent with reports from experienced users who conjecture that those encoders achieve a very good trade-off between speed and precision. Even though this method is not theoretically founded as a means to tell encoder properties apart, the results look plausible.

The approach demonstrated in this section can even be generalised to a broader scope. We infer properties of algorithms from statistical distances between their output and the output of other (reference) algorithms. This can be useful in situations where the algorithm under investigation is either a black box or intractable to analysis with formal methods. Both cases apply to some of the MP3 encoders in this study. It is conceivable that other cover generating or transformation processes can be handled similarly.

## 7.6 Summary and Discussion

In this chapter, a method has been presented to determine the encoder of ISO/MPEG 1 Audio Layer-3 streams on the basis of statistical features extracted from the stream. We have developed a set of ten features and used it in conjunction with a naïve Bayes classifier to discriminate between 20 different MP3 encoders. The results indicate that the proposed method is quite reliable for our test data. However, we are fully aware that the proposed method is in no way 'optimal' and may need further refinement for real-world applications.

### 7.6.1 Limitations and Future Directions

The first obstacle is the relatively narrow range of supported bit rates. In order to keep the test database operable, we decided to concentrate on the most widely used bit rates. To mitigate this restriction, we designed the features independently of the bit rate. This approach appears effective, as we do not have any problems when classifying variable bit rate (VBR) files despite never explicitly designing a feature for VBR data. However, as some encoders change the stereo model for different bit rates—especially for more extreme settings—further analyses of the robustness of the features against changes of the bit rate may increase the reliability of the classification.

As already mentioned, MP3 files support different stereo modes and most encoders offer a variety of options to fine-tune the encoding result. Since the test database always uses the (most likely) default settings and the presented features do not care about other encoding modes, sophisticated encoding parameters may cause false classifications. Hence, the influence of stereo modes and other encoding options is an open question for future research. In addition, some of the current features rely on file parameters (e.g., total file size) or precisely evaluate the beginning of a track (e.g., the initial silence). These features will fail if only fragments of a stream are to be classified.

Regarding the composition of encoders in the training set, we mainly cover open source encoders and the most widely used encoders from Fraunhofer and Xing. The selection of encoder implementations included in our analysis was not very systematic. Even if we are quite confident that additional software encoders can be added with moderate effort, we still have not examined the characteristics of hardware encoders which, for example, are included in portable digital audio recorders. The typical optimisations that are necessary to implement the MP3 encoding algorithm in DSP hardware might cause features of a different nature than those we exploit to differentiate between software encoders.

## 7.6.2 Transferability to Other Formats

The results on MP3 files show that encoder detection is feasible and has useful applications for steganalysis and related areas. Hence, it might be an interesting question as to whether the approach can be generalised with adapted features to other data formats.

Obviously, the MP3 format is a good candidate for encoder detection for two reasons. First, the popularity of the format, and thus the demand for encoders, developed a market for a couple of parallel developments in the late 1990s. Second, the inclusion of a psycho-acoustic model simplifies the task of feature discovery, because it leverages small numerical differences in the signal decomposition to measurable statistics, such as block type frequencies. From this point of view, MPEG 2 audio or MPEG 4 video seem to be promising formats for similar research. Other formats, for example, the popular JPEG image compression scheme, might be quite harder to classify. This format is less complicated—at least in the way it is used in the overwhelming majority of cases—and the Independent JPEG Group offers a standard implementation that is included in many applications ([111]; see also Sect. 2.6.4.1).

However, judging from our experience with MP3, we are confident that similar methods can be constructed for most complex standards that leave degrees of freedom for implementations. The experiments in Chapter 6 on the effect of different DCT methods already give an impression of what kind of differences can be expected between different implementations of JPEG compressors. If degrees of freedom increase with format complexity, we can even be quite optimistic for future formats. Some discoveries we made, for example, the block type signature of open source encoders, back our optimism: as long as programmers leave identifying traces, even by violating the standards, whether unintentionally or intentionally, classification will be feasible. It is an open question whether the iterative analytical task of feature discovery can be automated at some point in time.

## 7.6.3 Related Applications

Apart from the advances in steganalytic reliability, the proposed method may have applications in two further ways. From an academic point of view, the insights gained from the analysis of inter-encoder differences in MP3 files can be used to construct new steganographic algorithms. If we know the parameters that are treated differently by different encoders, we can consider them as indeterministic and modify them (carefully!) to convey steganographic semantics. Also, the design of watermarking algorithms, which are robust against MP3 compression with arbitrary encoders, might gain from detailed knowledge about encoder differences.

Last but not least, tools derived from this approach can be applied in multimedia forensics. Knowledge about the encoder of a suspicious file may lead to inferences about a possible creator. However, we must note that it is possible to forge any of the presented features, at least if some effort is made. So, the output of any of these classifiers is always a probabilistic indication and must never be considered as a reliable probative fact (cf. [20, 89]).

# Part III
# Synthesis

# Chapter 8
# General Discussion

## 8.1 Summary of Results

This book (and its underlying dissertation) are, to the best of our knowledge, the first comprehensive work focused on the role of covers in steganography and steganalysis, with an emphasis on the latter, and it contains a string of contributions. To structure the summary of important results, we distinguish them by the type of evidence we have to support them.

### 8.1.1 Results Based on Informal Arguments

Intuition produces weak evidence, but new insights often turn out useful for developing and establishing a structure of the field, and to pose more precise research questions.

Our main contribution in this respect is the clear distinction between *empirical* covers and *artificial* channels (first in Sect. 2.6.1). Based on this distinction, we have proposed a system to structure approaches to digital steganography not only by assumptions on the adversary model (analogously to cryptography), but also, on a second dimension, by assumptions on the cover. This system allows us to make statements on the achievable steganographic security as well as on capacity bounds. It thereby consolidates diverse results in the literature, some of which previously appeared contradictory (cf. Sect. 3.4.4).

The distinction between empirical and artificial covers can be further connected with the concept of different 'paradigms' in prior art (cf. Sect. 2.4), and we have argued that the core difference between the paradigms can be better explained in terms of our distinction. A similar link could be established to the information-theoretic discussion on the role of indeterminacy in cover signals (cf. Sect. 3.4.5).

A second main contribution is our argument that practical steganography, in the spirit of Simmons' [217] anecdote, should always be considered as empirical science (e.g., Chapter 1). To allow deductive reasoning despite the empirical nature of the problem, we have revisited and partly reformulated existing theoretical work to combine it with *principles of epistemology*. Epistemological reasons suggest that the distribution of natural covers must be accepted as incognisable. Therefore, it is necessary (and the best one can do) to formulate cover models as hypotheses on the cover distribution, which can be tested against empirical observations. So, the purpose of abstraction and model building in steganography is not only to reduce the dimensionality of cover signals to a tractable space, but also to deal with uncertainty about reality. The relative *quality* of cover models that are implied in steganographic systems or detectors, i.e., their *ability to predict reality*, determines the supremacy of steganography or steganalysis (cf. Sect. 3.2).

In response to the observation that natural cover sources emit very diverse covers, and this diversity is amplified by a large number of possibly combined preprocessing operations, we argue that *conditional cover models* are appropriate means to deal with heterogeneity between covers (cf. Sect. 3.3). Accordingly, we can understand cover distributions as mixtures of a large number of different sources, some of which in fact could be modelled more precisely *if* the condition that a specific suspect object is generated by a specific source is known to, say, the steganalyst. Conditional cover models allow us to divide the complicated problem of modelling general cover distributions into a number of more specific and simpler problems. This has implications on the architecture of practical implementations for steganalysis software, which can be structured in a modular way along the specific (conditional) modelling problems.

Another finding worth discussing here is the conceptual distinction between *computational* and *observability* *bounds* in complexity-theoretic analyses of steganographic security (cf. Sect. 3.4). This distinction directly follows from the empirical perspective, but it is not closely linked to heterogeneous empirical covers. So, the ultimate analysis of the relation between both bounds is left for future theoretical work.

Lastly, the unprecedented presentation of the foundations of steganography and steganalysis in a modular bottom-up manner with consistent terminology can be framed as a (minor) result and should be mentioned here (cf. Chapter 2).

### 8.1.2 Results Based on Mathematical Proofs

Owing to the empirical focus of this work, only a single result is based on a rigourous mathematical argument. We have shown that the coefficient structure with negative linear predictor signs for diagonal coefficients can in fact be derived from a simplified constant correlation image model (Sect. 6.1.1).

## 8.1.3 Results Based on Empirical Evidence

The lion's share of findings in Part II of this work is backed with empirical evidence.

In Chapter 4 we have proposed a *targeted detector* against the embedding function MB1. Experimental results suggest that the detector works very reliably for embedding rates above 50%, independently of the JPEG quality. Below 50%, detection is still possible in many cases of our test data, but we are careful and do not generalise this result too much. Nevertheless, the successful detection points to an unforeseen weakness of the cover model of MB1, and we have demonstrated how a superior cover model enables the steganalyst to unveil stego objects.

The study of heterogeneity between covers presented to quantitative detectors has led to the following three noteworthy results of Chapter 5:

1. Estimation errors in quantitative steganalysis are distributed according to a *heavy-tailed distribution*, which can be approximated reasonably well with a distribution of the Student $t$ family of distributions. This result calls into question the common practice of aggregating estimation errors with metrics based on moments of the Gaussian distribution (because these measurements may not converge for the true distribution). At the same time, nonparametric aggregate metrics as well as metrics based on parameters of fitted Student $t$ distributions can serve as viable alternatives.

2. It is possible to *decompose estimation errors* in quantitative steganalysis into (at least) one image-specific and two message-specific components. The only observable quantity in practice, however, is the compound error. Its message-specific component can be stripped off (or attenuated) by simulation experiments. This enables us to analyse in isolation the image-specific component, which is most relevant with regard to heterogeneity between covers. The relative magnitude of all error components has been measured and documented for five detectors under various conditions.

3. Approximating the estimation error with a Student $t$ distribution allows us to plug them as response variable in heteroscedastic Student $t$ regression models. This gives us a toolbox to model the *relation between macroscopic image properties and detection performance*, and to test for statistical significance. The individual and joint influence of two exemplary image properties, local variance and saturation, as well as the embedding rate as control variable, have been estimated and documented. Whenever influential macroscopic cover properties are available (or can be estimated) for a given suspect object, they can be used in conditional cover models to reduce heterogeneity and thus improve the accuracy of steganalysis.

Chapter 6 has dealt with *improvements of the weighted stego image (WS) analysis method.* On the one hand, a refinement of the method's explicit cover model, along with its weighting scheme, has lead to measurable and consistent performance improvements, which are robust (in relative terms) across heterogeneous conditions, such as varying preprocessing operations or replications on independent image sets (cf. Sect. 6.1). On the other hand, replacing the cover model with a conditional cover model for JPEG pre-compressed images yields substantial performance gains for this practically relevant class of (previously difficult to analyse with WS) covers (cf. Sect. 6.2).

Also the findings on audio steganalysis in Chapter 7 exploit the idea of conditional cover models. In this example, the challenge was to determine the condition by finding the (most likely) encoder of a given MP3 cover. The results generated with our proposed machine learning technique indicate that MP3 encoders leave sufficient identifying traces in their output to allow a very *reliable assignment of an observed file to its encoder* out of a set of 20 different encoders. Further, it is shown that such a pre-classification can improve steganalysis performance in practical scenarios, in which the heterogeneity between covers created with different encoders would otherwise push the error rates to unacceptable heights.

## 8.2 Limitations

As every empirical research is fallible, it is appropriate to recall the most important general limitations.[1]

Firstly and most importantly, the theory developed in Chapter 3 is not backed up *as a whole* with strong evidence. Rather, individual and selected examples in the Part II of this book have been used and presented in a way to illustrate selected aspects of the theory. This is definitely a shortcoming which has to be borne in mind when referring to the theory, but it also seems to be an unavoidable one. At least, we are not able to conceive a comprehensive test case for the theory (even when disregarding resource constraints which would prevent us from actually conducting the test).

Secondly, and important mainly to practitioners, this book contributes few advances to the perceived 'hard problems' in steganalysis, e.g., reliable detection of LSB matching. This has conceptional and computational reasons. When approaching new research problems, such as heterogeneity in cover signals, we believe it is sensible to first tackle the best understood cases before advancing to more obscure problems. Quantitative detection of LSB replacement in spatial domain images is probably *the best understood steganalysis problem* today. But, as can be seen from the discussion in the previous chapters, even this area leaves many puzzles and we are far from being able to

---

[1] Here, we do not repeat the specific limitation discussed in the chapters of Part II.

claim that the effect of covers on detection performance are sufficiently well understood *even in this simple scenario*. Given the empirical nature of the problem, we cannot expect that it will be fully understood at all. The aim was rather to propose general methods and concepts to deal with problems of heterogeneity, and to demonstrate their usefulness. So we are confident that the findings are transferrable to 'hard' problems once progress has been made and those problems are better understood. (We have no indication that heterogeneity of covers might be less relevant for the detection of LSB matching than of LSB replacement.) The computational reason refers to the fact that existing estimators of LSB matching steganography, despite being comparatively less reliable, are computationally much more demanding than LSB replacement estimators. This makes extensive simulations, such as the decomposition of estimation errors in Chapter 5, impractical with current technology.

All empirical research also demands a cautionary remark on the generalisability of such evidence. For historical reasons, the results of Chapters 4 and 7 were generated, by today's standards, from suboptimal and too homogeneous covers. We still deem the results relevant and informative, because the measurable effects are very clear and great care has been taken not to overfit the proposed detectors by abusive fine-tuning to our data set. Generalisability is a more critical issue for the results of Chapter 6, where much smaller performance differentials between alternative methods are interpreted. So it was necessary to validate these results with several independent and large ($N > 1500$) data sets originating from better-controlled sources. Nevertheless, whenever heterogeneity is suspected to 'drive' results, generalisation beyond the tested cover sources are most difficult. This is due to basic sampling theory: methods for statistical inference can in fact take into account (and mitigate) sampling and measurement errors, but are blind to systematic bias in the composition of the test data (coverage error). Since truly representative sets of natural images in communication channels relevant to steganography are not available, all steganography and steganalysis results are prone to this type of error. Despite these shortcomings, we believe that our research, both what is presented in this book and our recent conference publications, belong to the most (self-)critically tested results in the related literature.

## 8.3 Directions for Future Research

In the course of this book, we have touched upon numerous specific open research questions, virtually at every point where our analysis stopped. So we refrain from repeating all of them and just recall broader directions which we find relevant and promising.

### 8.3.1 Theoretical Challenges

Two important and, to the best of our knowledge, still unsolved theoretical aspects have emerged in Chapter 3. First, the role of the minimum sampling unit in complexity-theoretically secure steganography could be clarified with formal methods, and related to recent results on capacity. Also, it appears that the capacity reduction due to larger minimum sampling units can be traded off against higher embedding complexity, so it would be interesting to study bounds and optimal trade-offs. Second, and somewhat related, future research on complexity-theoretically secure steganography in empirical covers should scrutinise the distinction between the number of observations and the number of computations to clarify the relation between what we call observability bounds and classical computational bounds. As the problem does not appear unique to steganography, a review of similar problems in other areas could be a valuable first step.

### 8.3.2 Empirical Challenges

This work has interpreted heterogeneity in cover signals mainly as heterogeneity between cover objects. This is a reasonable starting point because this kind of heterogeneity, if not dealt with, is most susceptible to biasing summary measures of steganalysis performance. However, heterogeneity also exists on other levels: on a higher level between cover sources, and on a lower level between (groups of) samples within individual cover objects. The former has been touched upon in certain specific cases of this book, but a more comprehensive study (e.g., finding distribution models) is impeded by the problem that the source, i.e., the data set, becomes the unit of analysis, unlike the individual cover. Without additional simplifying assumptions, this squares the required sample size and simulation effort.

Heterogeneity on the sample level is not studied explicitly in this book, but our theoretical framework is general enough to work on all levels: for example, individual cover images could be interpreted as mixtures composed of areas of saturation, texture, noise, edges, etc. Analysing heterogeneity on this level might help to refine our understanding of nontrivial interactions between cover properties and detection performance, and creates a link to research on superposition models for images [221]. One example of such an interaction is the interdependence between areas of parity co-occurrence and smooth gradients in the choice of the weighting scheme for enhanced WS analysis (Sect. 6.1.2). Research in this direction could lead to steganalytic composition theorems or a sort of distributive law of detection performance. New insight might also be relevant for the construction of better-founded and thus more secure adaptive embedding operations.

### 8.3.3 Practical Challenges

Results from steganalysis research could become more valuable if considering heterogeneity between covers becomes a common practice. To facilitate this, large and freely available benchmark data sets from representative and heterogeneous sources should be made available to the community. Ideally these data sets should be annotated with metainformation on source and preprocessing properties. Also, a common methodology to draw reproducible samples from a larger repository of test data should be established. The most promising endeavour in this regard unfortunately was discontinued after a test phase [56].

Another practical challenge emerges when the idea of conditional cover models is taken to the extreme. As a result, the number of possible combinations of conditions and conditional models will explode and thus be hardly manageable with ad hoc techniques. A relevant open question is, to what extent can machine learning technologies help handle this complexity? Possible problems can be expected in the enormous data required to induce parameters from training data. Existing work on universal detectors based on machine learning techniques suggest that quite a bit of manual fine-tuning is still necessary to achieve good results on heterogeneous covers (e.g., heterogeneity due to different JPEG quality settings in [186, 190]).

## 8.4 Conclusion and Outlook

The evolution of cover models towards ever better approximations of true empirical covers has just begun and, given the number of combinations, will almost certainly go on for a while. Even the best-understood class of covers (greyscale images in the transformed domain) leaves plenty of open questions.

Nevertheless, we deem it appropriate to conclude this work with some thoughts on the consequences of a hypothetical convergence in the race for ever better cover models. We have argued throughout this book that refinements of cover models may help the development of both more secure embedding functions and better detectors. The 'winner' in a particular situation depends on the relative quality of cover models used by the steganographer and steganalyst. But what happens if new insight into real covers is distributed symmetrically? In the long run, Kerckhoffs' [135] prediction materialises and we can expect that both parties are, on average, on the same technological level. This level will continue to rise with the passage of time as scientific evidence accumulates. The question that has remained open so far is about whether continuing progress on cover models ultimately helps the steganographer and the steganalyst equally, or whether, for some reason, one of the two benefits more than the other. In brief, do further discoveries asymptotically increase or decrease the achievable security of steganographic communication with empirical covers? Obviously, a general or theoretically rigourous answer to this question is beyond the scope of this concluding section. So we resort to informal arguments to justify our beliefs. The argument requires two assumptions:

1. Both steganographer and steganalyst learn about new discoveries at the same time and are always able to implement them flawlessly in their respective functions.
2. Scientific discoveries always concern new, possibly nonlinear, dependencies between parts of cover signals.

The second assumption can be justified by the epistemological fact that independence between two empirical phenomena cannot be 'validated' with finite observations (in other words, independence is always the null hypothesis to be falsified). We believe that steganographer and steganalyst are affected as follows:

- For the steganographer, if newly discovered dependencies are nonlinear, and thus cannot be inverted to derive an efficient synthesis method from independent random variables,[2] then either the dependencies are violated (which risks detection and must be avoided) or the embedding complexity increases (because combinations in line with the dependence relation have to be found by search algorithms).
- For the steganalyst, testing the existence of higher-order dependencies is not necessarily much more complex, but to maintain the power of the statistical test, larger sample sizes are needed.

Altogether, we conjecture that scientific progress in the search for better cover models, if equally available and adopted by steganographer and steganalyst, ceteris paribus implies increasing *computational complexity for the steganographer* and increasing *observational complexity for the steganalyst*. So the ultimate answer on who benefits more from scientific discovery depends on the relation between the two types of complexity, e.g., in terms of relative cost. This, again, emphasises the need to better understand both types of complexity. The conjecture is limited to empirical covers and contrasts with a recent theoretical result by Wang and Moulin [231], who find positive secure capacity if the cover distribution is perfectly known, as in artificial channels. This highlights how essential the conceptual distinction between these two classes of covers actually is.

Other commonly stated options for the steganographer to cope with better steganalysis are generally lower, and asymptotically decreasing [124], embedding rates or the cultivation of artificial channels, i.e., changing the conventions to make sending controlled (and well-modelled) indeterminacy more plausible [45]. The latter not only tweaks with the plausibility heuristic, but

---

[2] Linear correlation in multivariate Gaussian signals can be expressed as the weighted sum of independent random variables (see Sect. 6.1.1), but this is an exception and does not apply in general.

also breaks with our definition of the steganographic system. If the channel distribution is regarded as something (partly) under the engineer's control, then it moves from the outside of the system (cf. Sect. 1.1) to its inside. This attempt can be best described by the term 'social engineering' in its literal meaning, but at the same time poses a different problem than the one studied in this book.

Finally, let us open the scope beyond steganography and reflect on related areas in which part of our findings might be applicable. Typical candidates are areas where undetectability is a relevant protection goal. This includes multimedia forensics, in particular *tamper hiding*: these techniques aim to change the semantic of media data without leaving detectable traces of manipulation. Here, requirements, empirical nature, and the need for cover models correspond to the situation in steganography and steganalysis [20, 141]. Some of our insights might also be relevant for privacy-enhancing technologies based on the data avoidance principle. For example, the definition of *unlinkability* in [192] requires a relation to be indistinguishable from all other possible relations between two sets of entities. The way unlinkability is achieved in certain applications resembles our concept of artificial channels, i.e., the system ensures that all observable outputs comply with a defined (usually uniform) distribution. But other applications have a clear empirical dimension, e.g., whenever behavioural or physical aspects are involved that cannot be efficiently transformed to uniform distributions (location privacy, radio signatures, etc.). Nevertheless, the state of the art in this area is still largely confined to theoretical 'world models'. The problem of high dimensionality is acknowledged, but aspects such as dimension reduction by models, empirical distributions and (hardly avoidable) heterogeneity, are rarely considered, yet. So it is likely that some of the concepts stated in this book in the context of steganography and steganalysis can be reformulated in an unlinkability context and might turn out useful in the area of privacy-enhancing technologies.

# Appendix A
# Description of Covers Used in the Experiments

The following image sets have been used throughout this book:

- **Image set A: Raw camera images ('baseline set')** provided by Andrew Ker [133]. $N = 1,600$ images were obtained from a single Minolta DiMAGE A1 camera. All images were stored in raw format and extracted as 12-bit greyscale bitmaps, without any colour filter array interpolation or denoising. The size has been adjusted to exactly $2,000 \times 1,500$ by slight cropping. Operable sizes for steganalysis benchmarking were generated by controlled downsampling, typically to $640 \times 480$. Andrew Ker has offered to make these images available to other researchers upon request.

- **Image set B: Scanned NRCS images ('validation set')** downloaded from the National Resources Conservation Service (NRCS) photo gallery website [175]. $N = 800$ uncompressed eight-bit true-colour images in TIFF format, apparently scanned from film, drawn randomly from roughly $3,000$ images. Original size images of approximately $2,100 \times 1,500$ pixels have been converted to grey scale and then downsampled to operable sizes, typically $640 \times 457$. The preprocessing history of these images is less under the researcher's control, but we decided to use this source as a second independent 'validation set', as it has previously been used to benchmark steganography and therefore allows for a certain degree of comparability of results. All images are publicly available on the Internet. The list of sampled file names is available from the author upon request.

- **Image set C: Raw camera images ('exploration set')** provided by Hany Farid and Mikah Johnson [56]. $N \approx 300$ uncompressed eight-bit true-colour images in TIFF format taken as raw images from multiple digital cameras (and thus varying size between five and six megapixels) have been downloaded from Dartmouth College's Digital Forensic Image Library (DFIL) project, which unfortunately never went beyond its beta phase. Due to the small number of images and the suspension of the database project, we have not used these images for quantitative results presented in this book. However, selected images of this set, converted to grey scale and downsampled to operable sizes $< 1$ megapixel, have been used to test steganographic algorithms and detectors, and to calibrate parameters which have later been applied to independent validation sets. As of October 2009, the copyright situation for these images is unclear, and so is their availability.

- **Image set D: JPEG pre-compressed camera images ('convenience set')** taken in 2002 by the author with a Sony Cybershot DSC-F55E digital camera and stored as JPEG with highest possible quality. $N \approx 300$ images have been downsampled from $1,600 \times 1,200$ to $800 \times 600$ pixels, then re-compressed as JPEG. Their luminance channels were used in the steganalysis benchmarks of the MB1 detector in Chapter 4.
- **Image set E: Preprocessed camera images ('van Hateren set')** downloaded from Hans van Hateren's resource page [100]. These images have become a quasi-standard in the literature on mathematical studies of image properties, e.g., [221]. $N = 200$ allegedly 16-bit greyscale images ($1,536 \times 1,024$, unfiltered versions) have been used to identify spurious results reported for specific detectors of LSB matching due to the use of homogeneous and singular covers. The images are publicly available, but should not be used for steganalysis benchmarking for reasons given in Appendix B.

Descriptive statistics of image properties (cf. Sect. 5.2.1) of image sets A and B after various preprocessing chains are reported in Table A.1. Further, Table A.2 is useful for converting capacity units from JPEG to spatial domain image representations and vice versa.

Table A.1: Descriptive statistics of image properties

|  | Preprocessing | | | |
|---|---|---|---|---|
|  | bicubic | bilinear | n. n.[a] | cropped |
| **Image set A** | | | | |
| % of saturated pixels | 0.2 | 0.2 | 0.3 | 0.3 |
| % of images with saturation $> 5\%$ | 0.6 | 0.4 | 0.6 | 1.2 |
| average local variance[b] | 230.5 | 222.0 | 392.0 | 105.5 |
| **Image set B** | | | | |
| % of saturated pixels | 1.2 | 1.1 | 1.8 | 1.6 |
| % of images with saturation $> 5\%$ | 7.7 | 7.3 | 11.7 | 7.8 |
| average local variance[b] | 351.0 | 345.0 | 667.3 | 175.9 |

[a] nearest neighbour interpolation

[b] defined in Equation (5.16) on page 143

Table A.2: JPEG cover quantities for capacity calculation and unit conversion

| | **Unit: per pixel** | | | | | | **Unit: per bit of file size** | | | | | |
|---|---|---|---|---|---|---|---|---|---|---|---|---|
| | overall | | | | pre-proc. | | overall | | | | pre-proc. | |
| JPEG | quantile | | | | bicubic | n. n. | quantile | | | | bicubic | n. n. |
| quality $q$ | 0.25 | 0.50 | 0.75 | mean | mean | mean | 0.25 | 0.50 | 0.75 | mean | mean | mean |
| **Nonzero AC DCT coefficients** | | | | | | | | | | | | |
| 0.5 | 0.11 | 0.15 | 0.19 | 0.15 | 0.15 | 0.16 | 0.16 | 0.17 | 0.17 | 0.16 | 0.16 | 0.16 |
| 0.6 | 0.13 | 0.17 | 0.22 | 0.18 | 0.17 | 0.19 | 0.16 | 0.17 | 0.17 | 0.17 | 0.17 | 0.17 |
| 0.7 | 0.16 | 0.21 | 0.27 | 0.22 | 0.20 | 0.24 | 0.16 | 0.17 | 0.17 | 0.17 | 0.17 | 0.17 |
| 0.8 | 0.21 | 0.26 | 0.34 | 0.28 | 0.25 | 0.31 | 0.17 | 0.17 | 0.17 | 0.17 | 0.17 | 0.17 |
| 0.9 | 0.32 | 0.40 | 0.49 | 0.41 | 0.37 | 0.46 | 0.16 | 0.17 | 0.17 | 0.17 | 0.16 | 0.17 |
| 0.95 | 0.47 | 0.56 | 0.66 | 0.56 | 0.50 | 0.63 | 0.15 | 0.16 | 0.17 | 0.16 | 0.16 | 0.16 |
| 0.99 | 0.80 | 0.86 | 0.90 | 0.85 | 0.81 | 0.89 | 0.14 | 0.16 | 0.17 | 0.16 | 0.16 | 0.15 |
| **AC DCT coefficients of absolute value one** | | | | | | | | | | | | |
| 0.5 | 0.07 | 0.09 | 0.11 | 0.09 | 0.08 | 0.10 | 0.09 | 0.10 | 0.11 | 0.10 | 0.10 | 0.10 |
| 0.6 | 0.08 | 0.10 | 0.12 | 0.10 | 0.09 | 0.11 | 0.09 | 0.10 | 0.11 | 0.10 | 0.10 | 0.10 |
| 0.7 | 0.09 | 0.11 | 0.14 | 0.12 | 0.11 | 0.13 | 0.09 | 0.09 | 0.10 | 0.09 | 0.09 | 0.10 |
| 0.8 | 0.11 | 0.14 | 0.17 | 0.15 | 0.13 | 0.17 | 0.08 | 0.09 | 0.10 | 0.09 | 0.09 | 0.10 |
| 0.9 | 0.16 | 0.19 | 0.23 | 0.19 | 0.16 | 0.22 | 0.07 | 0.08 | 0.09 | 0.08 | 0.08 | 0.09 |
| 0.95 | 0.20 | 0.23 | 0.26 | 0.23 | 0.20 | 0.26 | 0.06 | 0.07 | 0.08 | 0.07 | 0.07 | 0.07 |
| 0.99 | 0.15 | 0.21 | 0.26 | 0.21 | 0.24 | 0.17 | 0.02 | 0.04 | 0.05 | 0.04 | 0.05 | 0.03 |

Statistics calculated from $N = 1600$ source covers (set A).

**Note:** The statement "the largest payload [secret message] that can be undetectably embedded in a JPEG file based on the current best blind steganalysis classifiers is about 0.05 bits per nonzero AC DCT coefficient" in [87] corresponds to an average secure capacity of $0.17 \times 0.05 = 0.85\%$ for the ratio of secret message length to compressed cover size (applicable to JPEG covers with $q = 0.7$).

# Appendix B
# Spurious Steganalysis Results Using the 'van Hateren' Image Database

The presumably spurious results of a novel detector of LSB matching proposed by Boncelet and Marvel [28] provide an apt example for the influence of image characteristics and the risk of benchmarking steganalysis with homogeneous image sets. Interestingly, at first sight, the authors conducted and documented their performance evaluation in a particularly exemplary manner: they tested their proposed detector, a machine learning technique fed with features calculated from bitplane compression rates, against large image databases (this was not so common until recently). Further, they drew on an existing and publicly available database of carefully compiled digital photographs, the so-called 'van Hateren' images [100]. This procedure is generally advisable, as it improves the reproducibility of results.

However, the surprisingly good performance against the otherwise hard-to-steganalyse LSB matching embedding operation caused suspicion. For example, Table 1 of [28] reports false positive rates as low as 0% (6%) at 50% detection rate for the detection of LSB replacement with embedding rates $p = 0.5$ ($p = 0.2$, respectively).

A closer look at the original description of the image preprocessing by van Hateren and van der Schaaf [100] supported our doubt.

> The image set consisted of 4212 images obtained with a Kodak DCS 420 digital camera (with a 28 mm camera lens). For the intensity this camera uses 12-bit sampling internally, which is then reduced to and stored as 8-bit data via a nonlinear scale table. As this table is recorded for each image, it can be used afterwards to expand the 8-bit data to a linear scale. Although the latter scale is, strictly speaking, not genuinely 12-bit deep, it is effectively close to it. [100, p. 360]

Also Boncelet and Marvel [28] report additional image processing to obtain eight-bit data, the typical cover format studied in steganalysis benchmarking:

> For testing and evaluation, we used 1200 images from the van Hateran database. These images are generally outdoor, nature images. They are greyscale and have never been compressed. The images are 1536 × 1024 pixels and were converted from 16 bits per pixel to 8 bits per pixel. Generally speaking, greyscale, never compressed, images are considered to be a difficult dataset for ±1 embedding steganalysis. [28, p. II-151]

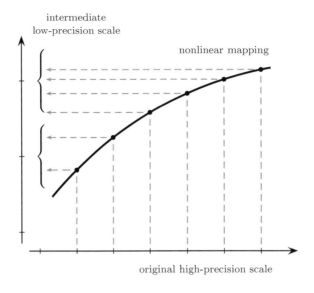

(a) Step 1

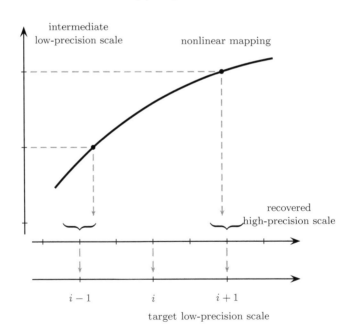

(b) Steps 2 and 3

Fig. B.1: Formation of singular histograms by repeated requantisation

So, effectively, the test images have been subject to a sequence of three requantisation operations on partly nonlinear scales. The impact of this transformation leaves singular traces in the image histogram, as can be seen from the schematic description of repeated requantisation in Figure B.1. In step 1, the original intensity values measured on a 12-bit scale are mapped along a concave mapping function to an eight-bit intermediate scale. The information loss materialises in the fact that multiple values on the high-precision scale are quantised to the same value on the low-precision scale (quantisation steps are depicted as curly brackets). Despite step 1 being not invertible, the values are mapped back to a high-precision scale in an attempt to recover the original 12-bit scale, though it is sparsely populated and "not genuinely 12-bit deep," as remarked in [100].

For the purpose of steganalysis benchmarking, quasi-12-bit images of the publicly available database have been yet again quantised (in a linear manner) to an eight-bit scale. This final scale contains singular artefacts, such as empty bins, as visualised for bin $i$ in Fig. B.1 (b). The formation of such artefacts is also visible in the scattered cover histogram of an example image depicted in Fig. B.2 (a). Both the peaks in the low intensity range of the histogram and the gaps in the high intensity range are due to sequential requantisation.

LSB matching, in its smoothing effect on image histograms, 'fills' such singular gaps and thus increases the entropy of the histogram (cf. Fig. B.2 (b)). It is obvious that measures of histogram-compressability are sensitive to this kind of change and thus allow reliable detection of LSB matching steganography—but only for such atypical covers!

Our suspicion was confirmed when we contacted the authors, who shared with us the results of a repeated performance evaluation with the NRCS images, a quasi-standard set for steganalysis benchmarks (the same as in our image set B). These results turned out to be much more plausible: for example, Table 1 of [168] reports a false positive rate of 5% at 80% detection rate, as opposed to 7.5% of the best-known universal steganalysis method against LSB replacement by Goljan et al. [91] (at embedding rate $p = 1$).

We may conclude that the 'van Hateren' images, which were originally obtained to study the human visual system [100], but also became a quasi-standard in the literature on general image models (e.g.,[221]), should not be used for benchmarking steganography or steganalysis. More generally, this is a warning example of over-interpreting results from singular or homogeneous image sets, the preprocessing of which is not fully under the researcher's control. We have to assume that this is not the only case of spurious results in the literature, and more subtle interdependencies of image micro-characteristics with embedding operations or detectors may be more difficult to spot, particularly if the cover sources are less well documented than in this case.

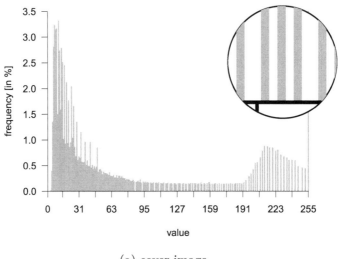

(a) cover image

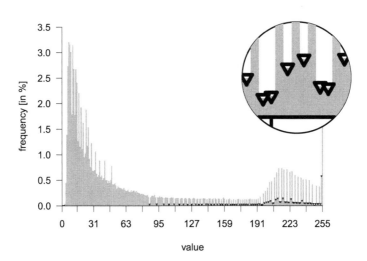

(b) stego image: PM1 with $p = 0.4$

Fig. B.2: Singular histogram due to sequential requantisation in the van Hateren image database and effect of LSB matching steganography. The 27 new nonzero bins in the stego image's histogram are highlighted with black marks

# Appendix C
# Proof of Weighted Stego Image (WS) Estimator

This proof first appeared in [73] and is adapted to our notation.

**Proposition 1.** *The net embedding rate $\hat{p}$ of a stego object $\boldsymbol{x}^{(p)}$ generated from cover $\boldsymbol{x}^{(0)}$ by changing approximately $\frac{1}{2}pn$ LSBs with the LSB replacement embedding operation can be estimated as two times the weight $\lambda$ that minimises the Euclidean distance between the weighted stego image $\boldsymbol{x}^{(p,\lambda)}$, defined as in Eq. (2.37), and the cover.*

*Proof.*

$$\hat{p} = 2 \arg\min_{\lambda} \sum_{i=1}^{n} \left( x^{(p,\lambda)} - x^{(0)} \right)^2 \tag{C.1}$$

$$= 2 \arg\min_{\lambda} \sum_{i=1}^{n} \left( \lambda(\overline{x}^{(p)} - x^{(p)}) + x^{(p)} - x^{(0)} \right)^2 \tag{C.2}$$

$$= 2 \arg\min_{\lambda} \sum_{x^{(p)}=x^{(0)}} \lambda^2 + \sum_{x^{(p)}\neq x^{(0)}} (\lambda - 1)^2 \tag{C.3}$$

$$\approx 2 \arg\min_{\lambda} \left( 1 - \frac{p}{2} \right) \lambda^2 + \frac{p}{2}(\lambda - 1)^2 \tag{C.4}$$

$$= 2 \arg\min_{\lambda} \left( 1 - \frac{p}{2} \right) \lambda^2 + \frac{p}{2}\lambda^2 - \lambda p + \frac{p}{2} \tag{C.5}$$

$$= 2 \arg\min_{\lambda} \lambda^2 - \lambda p + \frac{p}{2} \quad = \quad p \tag{C.6}$$

**Comments:** Equation (C.2) is obtained by inserting the definition of the weighted stego image from Eq. (2.37). Equation (C.3) captures the logic of the LSB replacement embedding operation (cf. Eq. 2.8). The sums can be replaced in (C.4) by the expected number of embedding changes from the definition of the net embedding rate $p$ and the number of samples $n$. Equation (C.5) and the left part of Eq. (C.6) are simple rearrangements. The first-order condition of the left part of Eq. (C.6) is $2\lambda - p = 0$, so the term reaches its unique minimum at $\lambda = \frac{1}{2}p$. The denominator cancels out with the leading factor 2.

# Appendix D
# Derivation of Linear Predictor for Enhanced WS

Let $\boldsymbol{\Sigma}$ be a correlation matrix of the form

$$\boldsymbol{\Sigma} = \begin{bmatrix} 1 & \varrho & \varrho & 0 \\ \varrho & 1 & 0 & \varrho \\ \varrho & 0 & 1 & \varrho \\ 0 & \varrho & \varrho & 1 \end{bmatrix}; \tag{D.1}$$

then, the Cholesky decomposition $\boldsymbol{a}$ that fulfils $\boldsymbol{\Sigma} = \boldsymbol{a}^\mathsf{T}\boldsymbol{a}$ is given as

$$\boldsymbol{a} = \begin{bmatrix} 1 & \varrho & \varrho & 0 \\ 0 & \left(1 - \varrho^2\right)^{\frac{1}{2}} & -\varrho^2 \left(1 - \varrho^2\right)^{-\frac{1}{2}} & \varrho \left(1 - \varrho^2\right)^{-\frac{1}{2}} \\ 0 & 0 & \left(1 - 2\varrho^2\right)^{\frac{1}{2}} \left(1 - \varrho^2\right)^{-\frac{1}{2}} & \varrho \left(2\varrho^4 - 3\varrho^2 + 1\right)^{-\frac{1}{2}} \\ 0 & 0 & 0 & \left(1 - 4\varrho^2\right)^{\frac{1}{2}} \left(1 - 2\varrho^2\right)^{-\frac{1}{2}} \end{bmatrix}. \tag{D.2}$$

Hence, the inverse of the top-left $3 \times 3$ sub-matrix $\boldsymbol{a}_\diamond$ is

$$\boldsymbol{a}_\diamond^{-1} = \begin{bmatrix} 1 & -\varrho \left(1 - \varrho^2\right)^{-\frac{1}{2}} & -\varrho \left(2\varrho^4 - 3\varrho^2 + 1\right)^{-\frac{1}{2}} \\ 0 & \left(1 - \varrho^2\right)^{-\frac{1}{2}} & \varrho^2 \left(2\varrho^4 - 3\varrho^2 + 1\right)^{-\frac{1}{2}} \\ 0 & 0 & \left(1 - \varrho^2\right)^{\frac{1}{2}} \left(1 - 2\varrho^2\right)^{-\frac{1}{2}} \end{bmatrix}; \tag{D.3}$$

so we obtain the following expressions for the coefficient $\boldsymbol{b}$ from Eq. (6.14):

$$
\boldsymbol{b} = \boldsymbol{a}_\diamond^{-1}\,\boldsymbol{v} =
\begin{bmatrix}
-\varrho(1-\varrho^2)^{-\frac{1}{2}} & -\varrho(2\varrho^4 - 3\varrho^2 + 1)^{-\frac{1}{2}} \\[4pt]
(1-\varrho^2)^{-\frac{1}{2}} & \varrho^2(2\varrho^4 - 3\varrho^2 + 1)^{-\frac{1}{2}} \\[4pt]
0 & (1-\varrho^2)^{\frac{1}{2}}(1-2\varrho^2)^{-\frac{1}{2}}
\end{bmatrix}
\begin{bmatrix}
\varrho(1-\varrho^2)^{-\frac{1}{2}} \\[4pt]
\varrho(2\varrho^4 - 3\varrho^2 + 1)^{-\frac{1}{2}}
\end{bmatrix}.
$$

$$\text{(D.4)}$$

Since $b_1 = b_2$ for symmetry, it is sufficient to regard only two rows:

$$
\begin{bmatrix} b_0 \\ b_1 \end{bmatrix} =
\begin{bmatrix}
\dfrac{\varrho^2}{\varrho^2 - 1} - \dfrac{\varrho^2}{2\varrho^4 - 3\varrho^2 + 1} \\[12pt]
-\dfrac{\varrho}{\varrho^2 - 1} + \dfrac{\varrho^3}{2\varrho^4 - 3\varrho^2 + 1}
\end{bmatrix}.
$$

$$\text{(D.5)}$$

After factoring the denominators, using $2\varrho^4 - 3\varrho^2 + 1 = (\varrho^2 - 1)(2\varrho^2 - 1)$, the expressions in Eqs. (6.15) and (6.16) follow from straight simplification.

# Appendix E
# Game for Formal Security Analysis

This game first appeared in [117] and is adapted to our notation. We stick close to its original formulation[1] despite some necessary refinements.

This interactive game between a steganalyst and a judge defines an adversary model for steganographic security that is independent from 1) knowledge of a 'true' probability distribution of covers $\mathsf{Prob}(\boldsymbol{x}^{(i)}|i = 0) = \mathcal{P}_0$ and 2) a prior on the ratio of stego objects to overall communication in reality $\mathsf{Prob}(i = 0)$ (symbols as in Eq. 3.4, p. 82).

It is assumed that the players have access to two oracles:

1. a *cover generating oracle* Sample : $\varnothing \to \mathcal{X}^*$ that returns a cover from an infinite sequence of covers (similarly to the observation of ordinary communication channels);
2. a so-called *structure evaluation oracle*, which corresponds to function Embed with a predefined key. Access to this oracle does not imply knowledge of the key.

The game is executed as follows:

- **Step 1:** The judge randomly picks a stego key $\boldsymbol{k} \in \mathcal{K}$ and gives the steganalyst a structure evaluation oracle $\mathsf{Embed}_{\boldsymbol{k}}$.
- **Step 2:** The steganalyst performs polynomial computations. During these computations it is allowed to query oracle $\mathsf{Embed}_{\boldsymbol{k}}$ with $N_1$ arbitrary messages $\boldsymbol{m}_1, \ldots, \boldsymbol{m}_{N_1}$ and covers $\boldsymbol{x}_1^{(0)}, \ldots, \boldsymbol{x}_{N_1}^{(0)}$, thus obtaining the corresponding stego objects $\boldsymbol{x}_1^{(1)}, \ldots, \boldsymbol{x}_{N_1}^{(1)}$ satisfying $\mathsf{Embed}(\boldsymbol{m}_i, \boldsymbol{x}_i^{(0)}, \boldsymbol{k}) = \boldsymbol{x}_i^{(1)}$ and $\mathsf{Extract}(\boldsymbol{x}_i^{(1)}, \boldsymbol{k}) = \boldsymbol{m}_i$ for $1 \leq i \leq N_1$. Furthermore, the steganalyst queries oracle Sample exactly $N_2$ times to obtain covers $\boldsymbol{x}_{N_1+1}^{(0)}, \ldots, \boldsymbol{x}_{N_1+N_2}^{(0)}$. All oracle queries can be interwoven and the input of one query can be

---

[1] One obvious problem in step 3 has been resolved to avoid unnecessary confusion: the original publication refers to $\boldsymbol{x}_1^{(0)}$ instead of $\boldsymbol{x}_{N_2+1}^{(0)}$, which would render the steganalyst's problem trivial.

dependent on the output of the previous oracle queries. The number of the oracle queries $N_1$ and $N_2$ is not restricted; the only requirement is that the total computation time spent on the game be polynomial. Note that the input to oracle $\mathsf{Embed}_k$ does not need to be generated by oracle $\mathsf{Sample}$ (i.e., the steganalyst can query $\mathsf{Embed}_k$ with pathologic covers, such as constant sample values).

- **Step 3:** After the steganalyst has finished the reasoning process, the judge selects two covers $\boldsymbol{x}_{N_2+1}^{(0)}$ and $\boldsymbol{x}_{N_2+2}^{(0)} \in \mathcal{X}^*$ by querying $\mathsf{Sample}$ twice, selects a message $\boldsymbol{m}$ randomly and computes $\boldsymbol{x}_{N_2+2}^{(1)} = \mathsf{Embed}(\boldsymbol{m}, \boldsymbol{x}_{N_2+2}^{(0)}, \boldsymbol{k})$. The judge flips a coin and issues to the steganalyst either the cover $\boldsymbol{x}_{N_2+1}^{(0)}$ or the stego object $\boldsymbol{x}_{N_2+2}^{(1)}$.

- **Step 4:** The steganalyst performs a probabilistic test in an attempt to decide whether he was given the stego object $\boldsymbol{x}_{N_2+2}^{(1)}$ or the plain cover $\boldsymbol{x}_{N_2+1}^{(0)}$. The advantage for the steganalyst is the probability of a correct guess minus $1/2$.

- **Step 5:** The stego system is secure for oracle $\mathsf{Sample}$ if the advantage of the steganalyst is negligible.

# Appendix F
# Derivation of ROC Curves and AUC Metric for Example Cover Models

Consider the 'world model' of Sect. 3.2.3 and embedding function $\mathsf{Embed}_1$ (see Fig. 3.1).

*Cover model (b)*

The distribution of cover and stego objects as a function of the value of $\mathsf{Proj}_{(b)}$ is given as follows:

| $\mathsf{Proj}_{(b)}$ | 0 | 1 |
|---|---|---|
| $\mathrm{Prob}\left(i = 0\vert\mathsf{Proj}_{(b)}(\boldsymbol{x}^{(i)})\right)$ | 1 | $1/3$ |
| $\mathrm{Prob}\left(i = 1\vert\mathsf{Proj}_{(b)}(\boldsymbol{x}^{(i)})\right)$ | 0 | $2/3$ |

We define a detector $\mathsf{Detect}_{(b)} : \{0, 1\} \to \{\mathsf{cover}, \mathsf{stego}\}$, which takes $\mathsf{Proj}_{(b)}(\boldsymbol{x}^{(i)})$ as input. For input 0, the value 'cover' is deterministic. For input 1, the value is indeterministic and $\mathsf{Detect}_{(b)}$ returns 'cover' with probability $1 - \tau$ and 'stego' with probability $\tau \in [0, 1]$.

As by definition $\mathrm{Prob}\left(\mathsf{Proj}_{(b)}(\boldsymbol{x}^{(0)}) = k\right) = \mathrm{const}\ \forall k \in \{0, 1\}$, the error rates $\alpha$ and $\beta$ can be written as functions of $\tau$,

$$\alpha = \frac{\tau}{2} \quad \text{and} \quad \beta = 1 - \tau. \tag{F.1}$$

A function for the shape of the ROC curve can be obtained by eliminating $\tau$ and expression $1 - \beta$ as a function of $\alpha$. The case differentiation is due to the constraint $\tau \le 1$ and Eq. (F.1).

$$\mathsf{ROC}_{(b)}(\alpha) = 1 - \beta = \begin{cases} 2\alpha \text{ for } \alpha \le \frac{1}{2} \\ 1 \quad \text{otherwise} \end{cases} \tag{F.2}$$

Integrating over $\alpha$ yields the area under the curve (AUC) metric,

$$\mathrm{AUC}_{(b)} = 2 \int_0^1 \mathrm{ROC}_{(b)}(\alpha) \, d\alpha - 1 \quad = \quad \frac{1}{2} \, . \tag{F.3}$$

*Cover model (c)*

The distribution of cover and stego objects as a function of the value of $\mathsf{Proj}_{(c)}$ is given as follows:

| $\mathsf{Proj}_{(c)}$ | 1 | 2 | 3 |
|---|---|---|---|
| Prob $\left( i = 0 \middle| \mathsf{Proj}_{(c)}(\boldsymbol{x}^{(i)}) \right)$ | 1 | 1/2 | 0 |
| Prob $\left( i = 1 \middle| \mathsf{Proj}_{(c)}(\boldsymbol{x}^{(i)}) \right)$ | 0 | 1/2 | 1 |

Detector $\mathsf{Detect}_{(c)}$ is deterministic for inputs 1 and 3 returning 'cover' and 'stego', respectively. Its value is indeterministic for input 2. Let $\tau$ again be the probability of output 'stego'. Using $\mathsf{Prob}\left( \mathsf{Proj}_{(c)}(\boldsymbol{x}^{(0)}) = k \right) = \mathrm{const} \; \forall k \in \{1,2\}$ and $\mathsf{Prob}\left( \mathsf{Proj}_{(c)}(\boldsymbol{x}^{(1)}) = k \right) = \mathrm{const} \; \forall k \in \{2,3\}$, we obtain $\alpha = \frac{1}{2}\tau$, $\beta = \frac{1}{2}(1 - \tau)$, and hence

$$\mathsf{ROC}_{(c)}(\alpha) = 1 - \beta = \begin{cases} \alpha + \frac{1}{2} & \text{for } \alpha \leq \frac{1}{2} \\ 1 & \text{otherwise} \end{cases} \quad \text{and} \tag{F.4}$$

$$\mathrm{AUC}_{(c)} = 2 \int_0^1 \mathsf{ROC}_{(c)}(\alpha) \, d\alpha - 1 \quad = \quad \frac{3}{4} \, . \tag{F.5}$$

# Appendix G
# Supplementary Figures and Tables

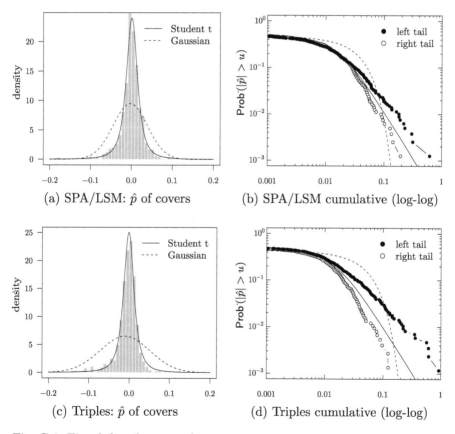

(a) SPA/LSM: $\hat{p}$ of covers

(b) SPA/LSM cumulative (log-log)

(c) Triples: $\hat{p}$ of covers

(d) Triples cumulative (log-log)

Fig. G.1: Fitted distributions of estimation error $\hat{p} - p$ for $p = 0$; data from 800 never-compressed greyscale images (set B); location and scale parameters estimated; $\nu = 2$

Table G.1: Regression coefficients fitted to $Z_{\text{cover}}$ of WS analysis

| Predictor | $\langle 1 \rangle$ | $\langle 2 \rangle$ | $\langle 3 \rangle$ | $\langle 4 \rangle$ | $\langle 5 \rangle$ | $\langle 6 \rangle$ | $\langle 7 \rangle$ |
|---|---|---|---|---|---|---|---|
| | | | | Specification | | | |
| **Location model** | | | | | | | |
| constant | 0.12 * (0.061) | −0.11 (0.310) | 1.47 *** (0.328) | 1.63 *** (0.461) | 0.66 *** (0.095) | 0.35 (0.425) | 0.07 (0.627) |
| local variance (log) | | 0.04 (0.060) | | −0.03 (0.056) | | 0.05 (0.077) | 0.11 (0.119) |
| saturation (log) | | | 0.13 *** (0.031) | 0.13 *** (0.030) | | | |
| emb. rate $p$ | | | | | 1.50 *** (0.229) | 1.55 *** (0.228) | 2.32 (1.504) |
| loc. var. (log) × $p$ | | | | | | | −0.15 (0.280) |
| **Scale model** | | | | | | | |
| constant [a] | −8.64 (0.079) | −11.67 (0.584) | −7.04 (0.307) | −9.84 (0.685) | −8.35 (0.109) | −10.14 (0.593) | −11.68 (0.877) |
| local variance (log) | | 0.55 *** (0.106) | | 0.49 *** (0.106) | | 0.32 ** (0.106) | 0.60 *** (0.158) |
| saturation (log) | | | 0.17 *** (0.030) | 0.16 *** (0.031) | | | |
| emb. rate $p$ | | | | | −0.35 (0.281) | −0.24 (0.282) | 3.82 (2.034) |
| loc. var. (log) × $p$ | | | | | | | −0.75 * (0.369) |

$N = 800$ never-compressed grey scale images (set B); std. errors in brackets; coefficients of location model scaled to percentage points of embedding rate; $\nu = 2$; significance levels: *** $\leq 0.001$, ** $\leq 0.01$, * $\leq 0.05$.
[a] no significance test computed due to lack of null hypothesis

Table G.2: Regression coefficients fitted to $\log \hat{\sigma}_i$ of $Z_{\text{pos}}$ for WS analysis

| Predictor | $\langle 1 \rangle$ | $\langle 2 \rangle$ | $\langle 3 \rangle$ | $\langle 4 \rangle$ | $\langle 5 \rangle$ | $\langle 6 \rangle$ | $\langle 7 \rangle$ |
|---|---|---|---|---|---|---|---|
| | | | | Specification | | | |
| constant [a] | −5.78 (0.019) | −7.68 (0.120) | −5.88 (0.072) | −7.94 (0.140) | −5.81 (0.033) | −7.72 (0.153) | −7.46 (0.245) |
| local variance (log) | | 0.35 *** (0.022) | | 0.36 *** (0.022) | | 0.35 *** (0.027) | 0.30 *** (0.044) |
| saturation (log) | | | −0.01 (0.007) | −0.02 *** (0.006) | | | |
| emb. rate $p$ | | | | | 1.51 *** (0.064) | 1.55 *** (0.058) | 0.96 * (0.422) |
| loc. var. (log) × $p$ | | | | | | | 0.11 (0.077) |
| **R-squared (adj.)** | | 0.24 | 0.00 | 0.25 | 0.41 | 0.51 | 0.51 |

$N = 800$ never-compressed grey scale images (set B); std. errors in brackets; specifications $\langle 1 \rangle$–$\langle 4 \rangle$ fitted for $p = 0.05$; significance levels: *** $\leq 0.001$, ** $\leq 0.01$, * $\leq 0.05$.
[a] no significance test computed due to lack of null hypothesis

## Table G.3: Regression coefficients fitted to $Z_{\text{cover}}$ of SPA

| Predictor | | | | Specification | | | |
|---|---|---|---|---|---|---|---|
| | $\langle 1 \rangle$ | $\langle 2 \rangle$ | $\langle 3 \rangle$ | $\langle 4 \rangle$ | $\langle 5 \rangle$ | $\langle 6 \rangle$ | $\langle 7 \rangle$ |
| **Location model** | | | | | | | |
| *constant* | 0.16 ** (0.060) | −0.39 (0.312) | 1.52 *** (0.291) | 1.21 ** (0.444) | 0.13 (0.072) | −0.16 (0.153) | −0.33 (0.385) |
| local variance (log) | | 0.10 (0.060) | | 0.04 (0.059) | | 0.05 (0.027) | 0.09 (0.074) |
| saturation (log) | | | 0.13 *** (0.028) | 0.12 *** (0.027) | | | |
| emb. rate $p$ | | | | | −0.01 (0.113) | 0.02 (0.106) | 0.24 (0.637) |
| loc. var. (log) × $p$ | | | | | | | −0.05 (0.122) |
| **Scale model** | | | | | | | |
| *constant* [a] | −8.66 (0.079) | −11.58 (0.584) | −7.60 (0.307) | −10.30 (0.685) | −8.54 (0.109) | −11.72 (0.593) | −12.69 (0.877) |
| local variance (log) | | 0.53 *** (0.106) | | 0.47 *** (0.106) | | 0.57 *** (0.106) | 0.74 *** (0.158) |
| saturation (log) | | | 0.11 *** (0.030) | 0.10 *** (0.031) | | | |
| emb. rate $p$ | | | | | −3.49 *** (0.281) | −3.41 *** (0.282) | −0.51 (2.034) |
| loc. var. (log) × $p$ | | | | | | | −0.53 (0.369) |

$N = 800$ never-compressed grey scale images (set B); std. errors in brackets; coefficients of location model scaled to percentage points of embedding rate; $\nu = 2$; significance levels: *** $\leq 0.001$, ** $\leq 0.01$, * $\leq 0.05$.
[a] no significance test computed due to lack of null hypothesis

## Table G.4: Regression coefficients fitted to $\log \hat{\sigma}_i$ of $Z_{\text{pos}}$ for SPA

| Predictor | | | | Specification | | | |
|---|---|---|---|---|---|---|---|
| | $\langle 1 \rangle$ | $\langle 2 \rangle$ | $\langle 3 \rangle$ | $\langle 4 \rangle$ | $\langle 5 \rangle$ | $\langle 6 \rangle$ | $\langle 7 \rangle$ |
| *constant* [a] | −5.72 (0.018) | −7.84 (0.114) | −5.98 (0.071) | −8.29 (0.130) | −5.84 (0.028) | −7.47 (0.129) | −7.85 (0.206) |
| local variance (log) | | 0.39 *** (0.021) | | 0.40 *** (0.020) | | 0.29 *** (0.023) | 0.36 *** (0.037) |
| saturation (log) | | | −0.03 *** (0.007) | −0.04 *** (0.006) | | | |
| emb. rate $p$ | | | | | 2.32 *** (0.054) | 2.35 *** (0.049) | 3.20 *** (0.356) |
| loc. var. (log) × $p$ | | | | | | | −0.15 * (0.065) |
| **R-squared (adj.)** | | 0.31 | 0.02 | 0.34 | 0.70 | 0.75 | 0.75 |

$N = 800$ never-compressed grey scale images (set B); std. errors in brackets; specifications $\langle 1 \rangle$–$\langle 4 \rangle$ fitted for $p = 0.05$; significance levels: *** $\leq 0.001$, ** $\leq 0.01$, * $\leq 0.05$.
[a] no significance test computed due to lack of null hypothesis

Table G.5: Regression coefficients fitted to $Z_{cover}$ of SPA/LSM

| Predictor | ⟨1⟩ | ⟨2⟩ | ⟨3⟩ | ⟨4⟩ | ⟨5⟩ | ⟨6⟩ | ⟨7⟩ |
|---|---|---|---|---|---|---|---|
| **Location model** | | | | | | | |
| *constant* | 0.08 (0.067) | −0.05 (0.339) | 1.76 *** (0.329) | 2.08 *** (0.485) | 0.12 (0.082) | 0.65 * (0.265) | −0.12 (0.504) |
| local variance (log) | | 0.02 (0.066) | | −0.07 (0.063) | | −0.11 * (0.049) | 0.04 (0.097) |
| saturation (log) | | | 0.17 *** (0.031) | 0.16 *** (0.030) | | | |
| emb. rate $p$ | | | | | −1.03 *** (0.239) | −0.90 *** (0.227) | 1.72 (1.249) |
| loc. var. (log) × $p$ | | | | | | | −0.52 * (0.243) |
| **Scale model** | | | | | | | |
| *constant* [a] | −8.44 (0.079) | −11.54 (0.584) | −7.29 (0.307) | −10.21 (0.685) | −8.54 (0.107) | −12.02 (0.592) | −12.04 (0.863) |
| local variance (log) | | 0.56 *** (0.196) | | 0.51 *** (0.196) | | 0.63 *** (0.106) | 0.63 *** (0.155) |
| saturation (log) | | | 0.13 *** (0.030) | 0.12 *** (0.031) | | | |
| emb. rate $p$ | | | | | −1.30 *** (0.377) | −1.19 ** (0.378) | −1.67 (2.701) |
| loc. var. (log) × $p$ | | | | | | | 0.09 (0.489) |

$N = 800$ never-compressed grey scale images (set B); std. errors in brackets; coefficients of location model scaled to percentage points of embedding rate; $\nu = 2$; significance levels: *** $\leq 0.001$, ** $\leq 0.01$, * $\leq 0.05$.
[a] no significance test computed due to lack of null hypothesis

Table G.6: Regression coefficients fitted to $\log \hat{\sigma}_i$ of $Z_{pos}$ for SPA/LSM

| Predictor | ⟨1⟩ | ⟨2⟩ | ⟨3⟩ | ⟨4⟩ | ⟨5⟩ | ⟨6⟩ | ⟨7⟩ |
|---|---|---|---|---|---|---|---|
| *constant* [a] | −5.73 (0.019) | −7.32 (0.127) | −6.06 (0.072) | −7.83 (0.145) | −5.98 (0.030) | −7.46 (0.145) | −7.40 (0.228) |
| local variance (log) | | 0.29 *** (0.023) | | 0.31 *** (0.023) | | 0.27 *** (0.026) | 0.26 *** (0.041) |
| saturation (log) | | | −0.03 *** (0.007) | −0.04 *** (0.006) | | | |
| emb. rate $p$ | | | | | 3.19 *** (0.099) | 3.23 *** (0.093) | 3.01 *** (0.673) |
| loc. var. (log) × $p$ | | | | | | | 0.04 (0.122) |
| **R-squared (adj.)** | | 0.17 | 0.03 | 0.21 | 0.57 | 0.62 | 0.62 |

$N = 800$ never-compressed grey scale images (set B); std. errors in brackets; specifications ⟨1⟩–⟨4⟩ fitted for $p = 0.05$; significance levels: *** $\leq 0.001$, ** $\leq 0.01$, * $\leq 0.05$.
[a] no significance test computed due to lack of null hypothesis

## Table G.7: Regression coefficients fitted to $Z_{\text{cover}}$ of Triples

| Predictor | $\langle 1 \rangle$ | $\langle 2 \rangle$ | $\langle 3 \rangle$ | $\langle 4 \rangle$ | $\langle 5 \rangle$ | $\langle 6 \rangle$ | $\langle 7 \rangle$ |
|---|---|---|---|---|---|---|---|
| | | | | Specification | | | |
| **Location model** | | | | | | | |
| constant | −0.10 (0.064) | 0.65 (0.352) | 0.78 ** (0.263) | 1.75 *** (0.464) | −0.07 (0.086) | 0.81 * (0.340) | 1.04 (0.542) |
| local variance (log) | | −0.14 * (0.067) | | −0.18 ** (0.068) | | −0.16 * (0.063) | −0.21 * (0.103) |
| saturation (log) | | | 0.09 *** (0.026) | 0.09 *** (0.025) | | | |
| emb. rate $p$ | | | | | −0.96 (0.820) | −1.14 (0.788) | −3.77 (4.710) |
| loc. var. (log) $\times$ $p$ | | | | | | | 0.51 (0.904) |
| **Scale model** | | | | | | | |
| constant [a] | −8.52 (0.079) | −11.18 (0.584) | −8.26 (0.307) | −10.89 (0.685) | −8.61 (0.110) | −11.43 (0.592) | −11.73 (0.871) |
| local variance (log) | | 0.48 *** (0.106) | | 0.46 *** (0.106) | | 0.51 *** (0.106) | 0.57 *** (0.157) |
| saturation (log) | | | 0.03 (0.030) | 0.02 (0.031) | | | |
| emb. rate $p$ | | | | | −0.31 (1.072) | −0.05 (1.073) | 3.81 (7.696) |
| loc. var. (log) $\times$ $p$ | | | | | | | −0.71 (1.398) |

$N = 800$ never-compressed grey scale images (set B); std. errors in brackets; coefficients of location model scaled to percentage points of embedding rate; $\nu = 2$; significance levels: *** $\leq 0.001$, ** $\leq 0.01$, * $\leq 0.05$.
[a] no significance test computed due to lack of null hypothesis

## Table G.8: Regression coefficients fitted to $\log \hat{\sigma}_i$ of $Z_{\text{pos}}$ for Triples

| Predictor | $\langle 1 \rangle$ | $\langle 2 \rangle$ | $\langle 3 \rangle$ | $\langle 4 \rangle$ | $\langle 5 \rangle$ | $\langle 6 \rangle$ | $\langle 7 \rangle$ |
|---|---|---|---|---|---|---|---|
| | | | | Specification | | | |
| constant [a] | −5.47 (0.018) | −6.82 (0.124) | −5.91 (0.068) | −7.44 (0.140) | −6.07 (0.030) | −7.44 (0.133) | −7.13 (0.223) |
| local variance (log) | | 0.25 *** (0.022) | | 0.27 *** (0.022) | | 0.25 *** (0.024) | 0.19 *** (0.040) |
| saturation (log) | | | −0.04 *** (0.007) | −0.05 *** (0.006) | | | |
| emb. rate $p$ | | | | | 7.39 *** (0.264) | 7.54 *** (0.248) | 4.48 * (1.774) |
| loc. var. (log) $\times$ $p$ | | | | | | | 0.56 (0.323) |
| **R-squared (adj.)** | | 0.13 | 0.05 | 0.20 | 0.50 | 0.56 | 0.56 |

$N = 800$ never-compressed grey scale images (set B); std. errors in brackets; specifications $\langle 1 \rangle$–$\langle 4 \rangle$ fitted for $p = 0.05$; significance levels: *** $\leq 0.001$, ** $\leq 0.01$, * $\leq 0.05$.
[a] no significance test computed due to lack of null hypothesis

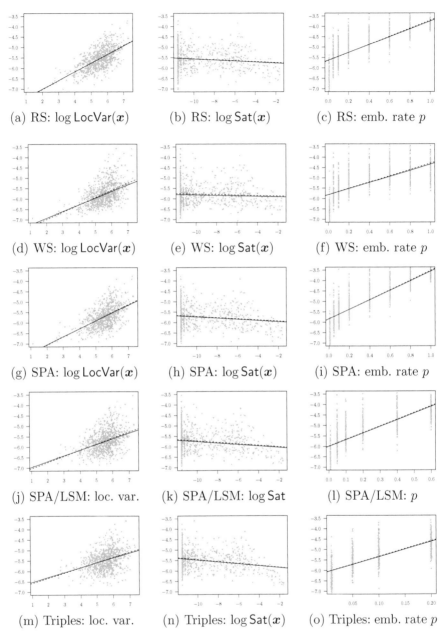

Fig. G.2: Bivariate relations with scale ($\log \hat{\sigma}$) of $Z_{\mathrm{pos}}$ as response ($y$-axis) in all curves; $p = 0.05$ except in the rightmost column; $N = 800$ never-compressed greyscale images (set B); OLS (dashed) and robust (solid) regression lines

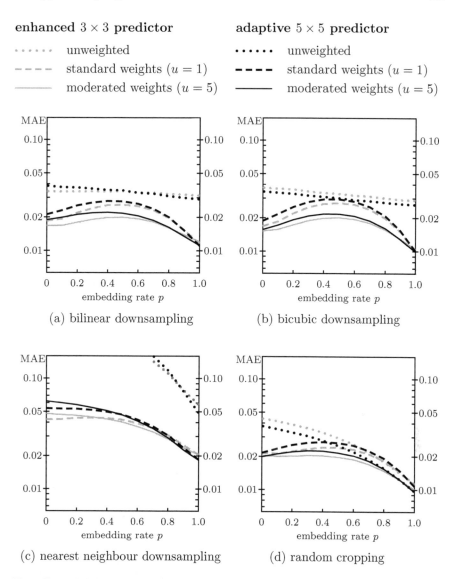

Fig. G.3: Validation of performance gains through WS improvements measured on image set B: MAE as a function of the embedding rate $p$ for different types of covers; smaller numbers indicate better performance; $N = 2,950$ scanned images downloaded from the NRCS database and reduced to $640 \times 457$; note the log scale

Table G.9: Performance indicators for enhanced WS on various cover sources

| indicator | MAE[a] | | | | IQR[b] | | | | median error[c] | | | |
|---|---|---|---|---|---|---|---|---|---|---|---|---|
| emb. rate $p$ | 0.0 | 0.1 | 0.4 | 1.0 | 0.0 | 0.1 | 0.4 | 1.0 | 0.0 | 0.1 | 0.4 | 1.0 |
| **bicubic down-sampling** | | | | | | | | | | | | |
| $3 \times 3$ pred., unw. | 2.4 | 2.3 | 2.2 | 1.9 | 3.0 | 2.9 | 2.9 | 2.7 | 0.5 | 0.4 | 0.3 | −0.0 |
| $3 \times 3$ pred., std. w. | 2.7 | 2.6 | 2.2 | 1.2 | 1.9 | 1.9 | 2.1 | 1.6 | 0.3 | 0.3 | 0.3 | 0.0 |
| $3 \times 3$ pred., mod. w. | 2.0 | 1.9 | 1.6 | 1.1 | 1.7 | 1.7 | 1.8 | 1.6 | 0.2 | 0.2 | 0.2 | 0.0 |
| $5 \times 5$ ad. pred., unw. | 2.1 | 2.1 | 2.0 | 1.7 | 2.8 | 2.8 | 2.7 | 2.6 | 0.3 | 0.2 | 0.2 | −0.0 |
| $5 \times 5$ ad. pred., std. w. | 2.9 | 2.8 | 2.4 | 1.2 | 1.9 | 2.0 | 2.1 | 1.7 | 0.3 | 0.3 | 0.4 | 0.0 |
| $5 \times 5$ ad. pred., mod. w. | 2.1 | 2.0 | 1.7 | 1.1 | 1.7 | 1.8 | 1.8 | 1.6 | 0.2 | 0.2 | 0.2 | 0.0 |
| $5 \times 5$ – ” –, bias corr. | 2.0 | 1.9 | 1.7 | 1.1 | 1.7 | 1.7 | 1.9 | 1.6 | 0.2 | 0.2 | 0.2 | 0.0 |
| Standard WS, unw. | 5.1 | 4.8 | 3.8 | 2.3 | 5.0 | 4.9 | 4.1 | 3.3 | 1.2 | 1.0 | 0.8 | 0.1 |
| Standard WS, w. | 2.9 | 2.8 | 2.4 | 1.3 | 2.0 | 2.1 | 2.1 | 1.6 | 0.3 | 0.3 | 0.4 | −0.0 |
| SPA[d] | 2.9 | 2.7 | 2.3 | 4.6 | 2.6 | 2.6 | 2.4 | 3.1 | 0.4 | 0.3 | 0.3 | −4.5 |
| **nearest neighbour down-sampling** | | | | | | | | | | | | |
| $3 \times 3$ pred., unw. | 13.0 | 11.9 | 8.7 | 3.9 | 10.1 | 8.8 | 7.3 | 5.5 | 2.5 | 2.4 | 1.6 | −0.3 |
| $3 \times 3$ pred., std. w. | 4.8 | 4.5 | 3.8 | 2.3 | 4.4 | 4.1 | 4.0 | 3.3 | 0.8 | 0.8 | 0.8 | 0.0 |
| $3 \times 3$ pred., mod. w. | 3.9 | 3.6 | 3.1 | 2.2 | 3.9 | 3.8 | 3.7 | 3.3 | 0.8 | 0.8 | 0.6 | 0.0 |
| $5 \times 5$ ad. pred., unw. | 15.8 | 14.4 | 10.2 | 3.4 | 10.7 | 9.9 | 7.6 | 4.8 | 2.7 | 2.8 | 2.0 | −0.2 |
| $5 \times 5$ ad. pred., std. w. | 4.7 | 4.4 | 3.6 | 2.2 | 4.0 | 3.9 | 3.7 | 3.1 | 0.7 | 0.7 | 0.6 | 0.0 |
| $5 \times 5$ ad. pred., mod. w. | 3.8 | 3.5 | 2.9 | 2.1 | 3.5 | 3.5 | 3.2 | 2.9 | 0.7 | 0.6 | 0.5 | −0.0s |
| $5 \times 5$ – ” –, bias corr. | 3.7 | 3.5 | 2.9 | 2.1 | 3.5 | 3.5 | 3.2 | 2.9 | 0.7 | 0.6 | 0.5 | −0.0 |
| Standard WS, unw. | 17.1 | 15.6 | 11.0 | 3.5 | 12.0 | 11.0 | 8.2 | 5.0 | 3.1 | 3.1 | 2.3 | −0.1 |
| Standard WS, w. | 4.8 | 4.5 | 3.8 | 2.7 | 4.4 | 4.3 | 4.1 | 3.5 | 0.8 | 0.9 | 0.7 | −0.0 |
| SPA[d] | 4.3 | 4.1 | 3.4 | 5.4 | 4.2 | 4.2 | 3.9 | 3.4 | 0.9 | 0.8 | 0.9 | −5.3 |
| **randomly cropped** | | | | | | | | | | | | |
| $3 \times 3$ pred., unw. | 2.6 | 2.4 | 2.0 | 1.4 | 2.3 | 2.3 | 2.2 | 1.9 | 0.4 | 0.4 | 0.4 | 0.0 |
| $3 \times 3$ pred., std. w. | 2.0 | 1.9 | 1.8 | 1.0 | 1.6 | 1.6 | 1.7 | 1.5 | 0.1 | 0.2 | 0.3 | −0.0 |
| $3 \times 3$ pred., mod. w. | 1.6 | 1.6 | 1.4 | 1.0 | 1.4 | 1.4 | 1.5 | 1.4 | 0.1 | 0.2 | 0.2 | −0.0 |
| $5 \times 5$ ad. pred., unw. | 1.8 | 1.7 | 1.5 | 1.2 | 1.8 | 1.8 | 1.8 | 1.7 | 0.1 | 0.1 | 0.1 | 0.0 |
| $5 \times 5$ ad. pred., std. w. | 2.0 | 1.9 | 1.7 | 0.9 | 1.4 | 1.4 | 1.6 | 1.4 | 0.1 | 0.1 | 0.2 | −0.0 |
| $5 \times 5$ ad. pred., mod. w. | 1.5 | 1.4 | 1.3 | 0.9 | 1.3 | 1.3 | 1.4 | 1.3 | 0.1 | 0.1 | 0.1 | 0.0 |
| $5 \times 5$ – ” –, bias corr. | 1.4 | 1.4 | 1.3 | 0.9 | 1.3 | 1.3 | 1.5 | 1.3 | 0.1 | 0.0 | 0.1 | 0.0 |
| Standard WS, unw. | 4.9 | 4.5 | 3.4 | 1.4 | 3.2 | 2.9 | 2.6 | 2.0 | 0.7 | 0.7 | 0.6 | 0.1 |
| Standard WS, w. | 2.3 | 2.3 | 2.1 | 1.3 | 1.9 | 2.0 | 2.1 | 1.8 | 0.2 | 0.3 | 0.4 | −0.0 |
| SPA[d] | 2.2 | 2.0 | 1.7 | 4.5 | 2.0 | 2.0 | 1.9 | 2.8 | 0.3 | 0.2 | 0.4 | −4.4 |

$N = 1,600$ raw camera images (set A) reduced to $640 \times 480$ pixels.
a) mean absolute error (in percentage points); a summary measure
b) inter-quartile range (in p. p.); a robust measure of dispersion
c) median error (in p. p.); a robust measure of bias
d) SPA failures omitted (see Footnote 7, p. 168)

## Table G.10: Performance of WS adopted to JPEG pre-compressed covers

| indicator | MAE[a] | | | | IQR[b] | | | | median error[c] | | | | FP$_{50}$[d] |
|---|---|---|---|---|---|---|---|---|---|---|---|---|---|
| emb. rate $p$ | 0.0 | 0.1 | 0.4 | 1.0 | 0.0 | 0.1 | 0.4 | 1.0 | 0.0 | 0.1 | 0.4 | 1.0 | 0.01 |
| **q = 0.5** | | | | | | | | | | | | | |
| JPEG WS, unw. | 0.1 | 0.1 | 0.1 | 0.2 | 0.0 | 0.1 | 0.2 | 0.3 | 0.0 | 0.0 | 0.0 | −0.0 | 0.6 |
| JPEG WS | 0.0 | 0.1 | 0.2 | 0.2 | 0.0 | 0.1 | 0.2 | 0.3 | 0.0 | 0.0 | 0.0 | 0.0 | 0.8 |
| enhanced WS | 1.0 | 3.5 | 9.6 | 0.6 | 1.0 | 4.1 | 12.5 | 0.7 | −0.3 | 3.1 | 9.2 | −0.1 | 5.3 |
| standard WS | 1.3 | 4.3 | 9.9 | 0.9 | 1.0 | 3.7 | 11.0 | 0.8 | 0.2 | 4.3 | 10.0 | −0.1 | 15.7 |
| SPA[e] | 1.5 | 1.4 | 1.1 | 3.9 | 1.1 | 1.1 | 0.9 | 2.5 | 0.3 | 0.3 | 0.2 | −3.6 | 22.8 |
| **q = 0.6** | | | | | | | | | | | | | |
| JPEG WS, unw. | 0.0 | 0.1 | 0.1 | 0.3 | 0.0 | 0.1 | 0.2 | 0.4 | 0.0 | 0.0 | 0.0 | −0.2 | 0.6 |
| JPEG WS | 0.0 | 0.1 | 0.2 | 0.5 | 0.0 | 0.2 | 0.3 | 0.6 | 0.0 | 0.0 | 0.0 | −0.3 | 0.6 |
| enhanced WS | 0.8 | 3.9 | 9.3 | 0.6 | 0.7 | 4.8 | 13.0 | 0.8 | 0.1 | 3.8 | 8.9 | −0.1 | 9.6 |
| standard WS | 1.5 | 4.1 | 9.1 | 1.0 | 1.5 | 3.8 | 10.9 | 0.9 | 0.3 | 3.9 | 8.9 | −0.1 | 18.7 |
| SPA[e] | 1.6 | 1.5 | 1.2 | 3.9 | 1.3 | 1.2 | 1.1 | 2.7 | 0.3 | 0.3 | 0.1 | −3.7 | 22.9 |
| **q = 0.7** | | | | | | | | | | | | | |
| JPEG WS, unw. | 0.0 | 0.1 | 0.9 | 0.4 | 0.0 | 0.1 | 1.3 | 0.5 | 0.0 | 0.0 | 0.2 | −0.3 | 0.4 |
| JPEG WS | 0.0 | 0.1 | 2.4 | 0.6 | 0.0 | 0.2 | 3.5 | 0.8 | 0.0 | 0.0 | 0.8 | −0.5 | 0.3 |
| enhanced WS | 0.7 | 3.3 | 8.1 | 0.7 | 0.6 | 4.4 | 12.0 | 0.9 | 0.1 | 3.0 | 7.2 | −0.1 | 7.6 |
| standard WS | 1.5 | 3.8 | 8.1 | 1.0 | 1.2 | 3.7 | 10.3 | 0.9 | 0.2 | 3.3 | 7.4 | −0.0 | 17.7 |
| SPA[e] | 1.7 | 1.5 | 1.3 | 4.0 | 1.2 | 1.2 | 1.1 | 2.8 | 0.3 | 0.2 | 0.1 | −3.7 | 22.6 |
| **q = 0.8** | | | | | | | | | | | | | |
| JPEG WS, unw. | 0.0 | 0.1 | 0.6 | 1.0 | 0.0 | 0.1 | 0.8 | 0.7 | 0.0 | 0.0 | 0.2 | −0.9 | 0.1 |
| JPEG WS | 0.0 | 0.1 | 1.6 | 1.4 | 0.0 | 0.2 | 2.2 | 1.3 | 0.0 | 0.0 | 0.5 | −1.2 | 0.3 |
| enhanced WS | 0.9 | 3.0 | 6.5 | 0.7 | 0.8 | 3.7 | 9.9 | 0.9 | 0.1 | 2.5 | 5.3 | −0.1 | 12.4 |
| standard WS | 1.8 | 3.5 | 6.8 | 1.2 | 1.5 | 3.2 | 8.5 | 1.0 | 0.2 | 2.8 | 5.9 | −0.1 | 20.5 |
| SPA[e] | 1.8 | 1.6 | 1.3 | 4.2 | 1.4 | 1.3 | 1.2 | 2.9 | 0.3 | 0.2 | 0.2 | −3.9 | 24.4 |
| **q = 0.95** | | | | | | | | | | | | | |
| JPEG WS, unw. | 0.1 | 1.1 | 8.2 | 30.5 | 0.1 | 0.2 | 0.3 | 0.5 | 0.0 | −1.1 | −8.2 | −30.5 | 0.4 |
| JPEG WS | 0.2 | 1.1 | 8.0 | 30.9 | 0.2 | 0.3 | 0.5 | 0.9 | 0.0 | −1.1 | −8.2 | −30.9 | 4.0 |
| enhanced WS | 1.3 | 1.6 | 2.3 | 1.0 | 1.3 | 1.6 | 2.4 | 1.3 | 0.2 | 0.6 | 1.0 | −0.1 | 19.9 |
| standard WS | 2.3 | 2.6 | 3.3 | 1.3 | 1.8 | 2.2 | 3.1 | 1.5 | 0.3 | 0.9 | 1.6 | −0.1 | 24.9 |
| SPA[e] | 2.0 | 1.9 | 1.6 | 4.4 | 1.8 | 1.7 | 1.6 | 2.9 | 0.3 | 0.3 | 0.2 | −4.2 | 26.0 |

$N = 1,600$ raw camera images (set A) down-sampled (bicubic) to $640 \times 480$ and then pre-compressed; weighted WS methods unless otherwise stated.
a) mean absolute error (in percentage points); a summary measure
b) inter-quartile range (in p. p.); a robust measure of dispersion
c) median error (in p. p.); a robust measure of bias
d) false positive rate at 50 % detection rate (in %)
e) SPA failures omitted (no real root of the estimation equation, see Footnote 7, p. 168)

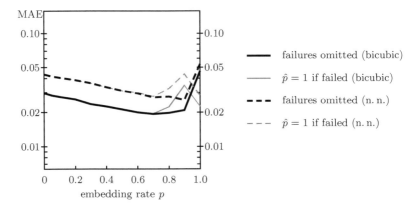

Fig. G.4: Impact of different treatment of SPA failures for $p$ close to 1; MAE of SPA detector for $N = 1,600$ images (set A) downsampled to $640 \times 480$

Table G.11: List of MP3 encoders included in the analysis of Chapter 7

| Mnemonic | Name | Author/Publisher | Verison | Year |
|---|---|---|---|---|
| 8hz-mp3 | 8HZ-MP3 Encoder | 8Hz Productions | 02b | 1998 |
| bladeenc | BladeEnc | T. Jansson | 0.94.2 | 2001 |
| fastenc | FastEnc | Fraunhofer IIS | 1.02 | 2000 |
| fhgprod | Fraunhofer MP3 Producer | Opticom | 2.1 | 1998 |
| gogo | gogo301 petit | Herumi and Pen | 3.01 | 2001 |
| iTunes | Apple iTunes | Apple Computer Inc. | 4.1-52 | 2003 |
| l3enc272 | l3enc (Linux) | Fraunhofer IIS | 2.72 | 1997 |
| l3encdos | l3enc (MS-DOS) | Fraunhofer IIS | 2.60 | 1996 |
| lame | LAME Ain't an MP3 Encoder | M. Cheng/M. Taylor et al. | 3.93 | 2003 |
| m3ec | M3E Command Line Version | N/A | 0.98b | 2000 |
| mp3comp | MP3 Compressor | MP3hC | 0.9f | 1997 |
| mp3enc31 | mp3enc (Demo) | Fraunhofer IIS | 3.1 | 1998 |
| plugger | Plugger | A. Demichelis | 0.4 | 1998 |
| shine | Shine | G. Bouvigne | 0.1.4 | 2001 |
| soloh | SoloH MPEG Encoder | N/A | 0.07a | 1998 |
| soundjam | SoundJam (Macintosh) | Casady and Greene | 2.5.1 | 2000 |
| uzura | Uzura | N/A (Fortran code) | 3.0 | 2002 |
| xing3 | Xing MP3 Encoder | Xing Technology Corp. | 3.0-32 | 1997 |
| xing98 | Xing MP3 Encoder (x3enc) | Xing Technology Corp. | 1.02 | 1998 |
| xingac21 | AudioCatalyst | Xing Technology Corp. | 2.10 | 1999 |

# References

[1] 8Hz Productions: 8hz-mp3 (1998). http://www.8hz.com/mp3/ (last access: October 2009)

[2] Abdalla, M., Bellare, M., Rogaway, P.: The oracle Diffie–Hellman assumptions and an analysis of DHIES. In: Naccache, D. (ed.) Topics in Cryptology. Proc. of CT-RSA, LNCS 2020, pp. 143–158. Springer-Verlag, Berlin Heidelberg (2001)

[3] Ahn, L.v., Hopper, N.J.: Public-key steganography. In: Cachin, C., Camenisch, J. (eds.) Proc. of EUROCRYPT, LNCS 3027, pp. 323–341. Springer-Verlag, Berlin Heidelberg (2004)

[4] Anderson, R.J.: Stretching the limits of steganography. In: Anderson, R.J. (ed.) Information Hiding (1st International Workshop), LNCS 1174, pp. 39–48. Springer-Verlag, Berlin Heidelberg (1996)

[5] Anderson, R.J., Petitcolas, F.A.P.: On the limits of steganography. IEEE Journal on Selected Areas in Communications **16**, 474–481 (1998)

[6] Arjun, S., Negi, A., Kranthi, C., Keerthi, D.: An approach to adaptive steganography based on matrix embedding. In: Proc. of IEEE TENCON (Region 10 Conference) (2007)

[7] Avcibaş, I., Memon, N.D., Sankur, B.: Steganalysis based on image quality metrics. In: Delp, E.J., Wong, P.W. (eds.) Security, Steganography and Watermarking of Multimedia Contents III (Proc. of SPIE), vol. 4314. San Jose, CA (2001)

[8] Backes, M., Cachin, C.: Public Key Steganography with Active Attacks. Report 2003/231. Cryptology ePrint Archive (2003). Online available at http://eprint.iacr.org/2003/231 (last access: October 2009)

[9] Barbier, J., Alt, S.: Practical insecurity for effective steganalysis. In: Solanki, K., Sullivan, K., Madhow, U. (eds.) Information Hiding, LNCS 5284, pp. 195–208. Springer-Verlag, Berlin Heidelberg (2008)

[10] Barbier, J., Filiol, É., Mayoura, K.: Universal detection of JPEG steganography. Journal of Multimedia **2**(2), 1–9 (2007)

[11] Bendens, O.: Geometry-based watermarking of 3D models. IEEE Computer Graphics and Applications **19**(1), 46–55 (1999)

[12] Benenson, Z., Kühn, U., Lucks, S.: Cryptographic attack metrics. In: Eusgeld, I., Freiling, F., Reussner, R. (eds.) Dependability Metrics, LNCS 4909, pp. 133–156. Springer-Verlag, Berlin Heidelberg (2008)

[13] Bergmair, R.: A comprehensive bibliography of linguistic steganography. In: Delp, E.J., Wong, P.W., Dittmann, J., Memon, N.D. (eds.) Security, Steganography and Watermarking of Multimedia Contents IX (Proc. of SPIE), vol. 6505. San Jose, CA (2007)

[14] Bierbrauer, J.: Crandall's problem. Mimeo (2001). Online available at http://www.ws.binghamton.edu/fridrich/covcodes.pdf (last access: October 2009)

[15] Böhme, R.: Assessment of steganalytic methods using multiple regression models. In: Barni, M., Herrera-Joancomartí, J., Katzenbeisser, S., Pérez-González, F. (eds.) Information Hiding (7th International Workshop), LNCS 3727, pp. 278–295. Springer-Verlag, Berlin Heidelberg (2005)

[16] Böhme, R.: Wet paper codes for public key steganography? Unpublished rump session talk at the 7th Information Hiding Workshop, Barcelona, Spain (2005). Presentation slides available at http://www.inf.tu-dresden.de/~rb21/publications/ Boehme2005_IHW_RumpSession.pdf (last access: October 2009)

[17] Böhme, R.: Improved Statistical Steganalysis using Models of Heterogeneous Cover Signals. Doctoral dissertation. Technische Universität Dresden, Faculty of Computer Science, Dresden, Germany (2008)

[18] Böhme, R.: Weighted stego-image steganalysis for JPEG covers. In: Solanki, K., Sullivan, K., Madhow, U. (eds.) Information Hiding, LNCS 5284, pp. 178–194. Springer-Verlag, Berlin Heidelberg (2008)

[19] Böhme, R.: An epistemological approach to steganography. In: Katzenbeisser, S., Sadeghi, A. (eds.) Information Hiding, LNCS 5806, pp. 15–30. Springer-Verlag, Berlin Heidelberg (2009)

[20] Böhme, R., Freiling, F.C., Gloe, T., Kirchner, M.: Multimedia forensics is not computer forensics. In: Geradts, Z.J.M.H., Franke, K., Veenman, C.J. (eds.) Proc. of International Workshop on Computational Forensics (IWCF), LNCS 5718, pp. 90–103. Springer-Verlag, Berlin Heidelberg (2009)

[21] Böhme, R., Keiler, C.: On the security of 'A steganographic scheme for secure communications based on the chaos and Euler theorem'. IEEE Transactions on Multimedia **9**, 1325–1329 (2007)

[22] Böhme, R., Ker, A.D.: A two-factor error model for quantitative steganalysis. In: Delp, E.J., Wong, P.W. (eds.) Security, Steganography and Watermarking of Multimedia Contents VIII (Proc. of SPIE), vol. 6072, pp. 59–74. San Jose, CA (2006)

[23] Böhme, R., Westfeld, A.: Breaking Cauchy model-based JPEG steganography with first order statistics. In: Samarati et al., P. (ed.) Computer Security (Proc. of ESORICS), LNCS 3193, pp. 125–140. Springer-Verlag, Berlin Heidelberg (2004)

[24] Böhme, R., Westfeld, A.: Exploiting preserved statistics for steganalysis. In: Fridrich, J. (ed.) Information Hiding (6th International Workshop), LNCS 3200, pp. 82–96. Springer-Verlag, Berlin Heidelberg (2004)

[25] Böhme, R., Westfeld, A.: Statistical characterisation of MP3 encoders for steganalysis. In: Proc. of ACM Multimedia and Security Workshop (MM&SEC), pp. 25–34. ACM Press, New York (2004)

[26] Böhme, R., Westfeld, A.: Feature-based encoder classification of compressed audio streams. ACM Multimedia Systems Journal 11(2), 108–120 (2005)

[27] Bollen, K.A.: Structural Equations with Latent Variables. Wiley, New York (1989)

[28] Boncelet, C.G., Marvel, L.M.: Steganalysis of $\pm 1$ embedding using lossless image compression. In: Proc. of IEEE ICIP, vol. 2, pp. 149–152 (2007)

[29] Bouchaud, J.P., Potters, M.: Theory of Financial Risk and Derivative Pricing: From Statistical Physics to Risk Management. second edn. Cambridge University Press (2004)

[30] Brandenburg, K., Stoll, G.: ISO-MPEG-1 audio: A generic standard for coding of high-quality digital audio. Journal of the Audio Engineering Society 42(10), 780–794 (1994)

[31] Brent, R.P., Gao, S., Lauder, A.G.B.: Random Krylov spaces over finite fields. SIAM Journal of Discrete Mathematics 16(2) (2003)

[32] Burges, C.J.C.: A tutorial on support vector machines for pattern recognition. Data Mining and Knowledge Discovery 2(2), 121–167 (1998)

[33] Cachin, C.: An information-theoretic model for steganography. In: Aucsmith, D. (ed.) Information Hiding (2nd International Workshop), LNCS 1525, pp. 306–318. Springer-Verlag, Berlin Heidelberg (1998)

[34] Cachin, C.: An information-theoretic model for steganography. Information and Computation 192, 41–56 (2004)

[35] Cancelli, G., Barni, M.: MPSteg-color: A new steganographic technique for color images. In: Furon, T., Cayre, F., Doërr, G., Bas, P. (eds.) Information Hiding, LNCS 4567, pp. 1–15. Springer-Verlag, Berlin Heidelberg (2007)

[36] Cancelli, G., Barni, M., Menegaz, G.: MPSteg: Hiding messages in the matching pursuit domain. In: Delp, E.J., Wong, P.W. (eds.) Security, Steganography and Watermarking of Multimedia Contents VIII (Proc. of SPIE), vol. 6072. San Jose, CA (2006)

[37] Chandramouli, R., Li, G., Memon, N.: Adaptive steganography. In: Wong, P.W., Delp, E.J. (eds.) Security, Steganography and Watermarking of Multimedia Contents IV (Proc. of SPIE), 4675, pp. 69–78. San Jose, CA (2002)

[38] Chandramouli, R., Memon, N.D.: Steganography capacity: A steganalysis perspective. In: Delp, E.J., Wong, P.W. (eds.) Security, Steganography and Watermarking of Multimedia Contents V (Proc. of SPIE), vol. 5022. San Jose, CA (2003)

[39] Chang, K.C., Chang, C.P., Huang, P.S., Tu, T.M.: A novel image steganographic method using tri-way pixel-value differencing. Journal of Multimedia **3**, 37–44 (2008)

[40] Chen, C., Shi, Y.Q., Xuan, G.: Steganalyzing texture images. In: Proc. of IEEE ICIP, vol. 2, pp. 153–156 (2007)

[41] Comesaña, P., Pérez-González, F.: On the capacity of stegosystems. In: Proc. of ACM Multimedia and Security Workshop (MM&SEC), pp. 15–24. Dallas, Texas, USA (2007)

[42] Cox, I., Miller, M., Bloom, J., Fridrich, J., Kalker, T.: Digital Watermarking and Steganography. 2nd edn. Morgan Kaufmann, Burlington, MA (2007)

[43] Crandall, R.: Some notes on steganography. Mimeo posted to a mailing list (1998). Online available at `http://os.inf.tu-dresden.de/~westfeld/crandall.pdf` (last access: October 2009)

[44] Craver, S.: On public-key steganography in the presence of an active warden. In: Aucsmith, D. (ed.) Information Hiding (2nd International Workshop), LNCS 1525, pp. 355–368. Springer-Verlag, Berlin Heidelberg (1998)

[45] Craver, S., Li, E., Yu, J., Atalki, I.: A supraliminal channel in a videoconferencing application. In: Solanki, K., Sullivan, K., Madhow, U. (eds.) Information Hiding, LNCS 5284, pp. 283–293. Springer-Verlag, Berlin Heidelberg (2008)

[46] Dabeer, O., Sullivan, K., Madhow, U., Chandrasekaran, S., Manjunath, B.S.: Detection of hiding in the least significant bit. IEEE Transactions on Signal Processing **52**(10), 3046–3058 (2004)

[47] Dempster, A., Laird, N., Rubin, D.: Maximum likelihood from incomplete data via the EM algorithm. Journal of the Royal Statistical Society, Series B (Methodological) **39**(1), 1–38 (1977)

[48] Draper, S.C., Ishwar, P., Molnar, D., Prabhakaran, V., Ramchandran, K., Schonberg, D., Wagner, D.: An analysis of empirical PMF based tests for least significant bit image steganography. In: Barni, M., Herrera-Joancomartí, J., Katzenbeisser, S., Pérez-González, F. (eds.) Information Hiding (7th International Workshop), LNCS 3727, pp. 327–341. Springer-Verlag, Berlin Heidelberg (2005)

[49] Duda, R.O., Hart, P.E.: Pattern Classification and Scene Analysis. Wiley, New York (1973)

[50] Dumitrescu, S., Wu, X., Wang, Z.: Detection of LSB steganography via sample pair analysis. IEEE Transactions on Signal Processing **51**, 1995–2007 (2003)

[51] Dumitrescu, S., Wu, X., Wang, Z.: Detection of LSB steganography via sample pair analysis. In: Petitcolas, F.A.P. (ed.) Information Hiding (5th International Workshop), LNCS 2578, pp. 355–372. Springer-Verlag, Berlin Heidelberg (2003)

[52] Eggers, J., Bäuml, R., Girod, B.: A communications approach to image steganography. In: Wong, P.W., Delp, E.J. (eds.) Security,

Steganography and Watermarking of Multimedia Contents IV (Proc. of SPIE), vol. 4675. San Jose, CA (2002)

[53] EncSpot – An MP3 Analyzer: (2002). http://www.mpex.net/ software/details/encspot.html (last access: October 2009)

[54] Ettinger, J.M.: Steganalysis and game equilibria. In: Aucsmith, D. (ed.) Information Hiding (2nd International Workshop), LNCS 1525, pp. 319–328. Springer-Verlag, Berlin Heidelberg (1998)

[55] Fan, Z., Queiroz, R.L.: Identification of bitmap compression history: JPEG detection and quantizer estimation. IEEE Transactions on Image Processing **12**(2), 230–235 (2003)

[56] Farid, H., Johnson, M.K.: The Digital Forensic Image Library (DFIL). Beta version at http://www.cs.dartmouth.edu/cgi-bin/cgiwrap/ isg/imagedb.py (last access: June 2005; not available in October 2009) (2004)

[57] Feig, E., Winograd, S.: On the multiplicative complexity of discrete cosine transforms. IEEE Transactions on Information Theory **38**(4), 1387–1391 (1992)

[58] Filler, T., Ker, A.D., Fridrich, J.: The square root law of steganographic capacity for Markov covers. In: Delp, E.J., Wong, P.W., Dittmann, J., Memon, N.D. (eds.) Media Forensics and Security XI (Proc. of SPIE), vol. 7254. San Jose, CA (2009)

[59] Fisher, R.A.: The use of multiple measurements in taxonomic problems. Annals of Eugenics **7**, 179–188 (1936)

[60] Fisk, G., Fisk, M., Papadopoulos, C., Neil, J.: Eliminating steganography in Internet traffic with active wardens. In: Petitcolas, F.A.P. (ed.) Information Hiding (5th International Workshop 2002), LNCS 2578, pp. 18–35. Springer-Verlag, Berlin Heidelberg (2003)

[61] Fontaine, C., Galand, F.: How can Reed-Solomon codes improve steganographic schemes? In: Furon, T., Cayre, F., Doërr, G., Bas, P. (eds.) Information Hiding, LNCS 4567, pp. 130–144. Springer-Verlag, Berlin Heidelberg (2007)

[62] Franz, E.: Steganography preserving statistical properties. In: Petitcolas, F.A.P. (ed.) Information Hiding (5th International Workshop 2002), LNCS 2578, pp. 278–294. Springer-Verlag, Berlin Heidelberg (2003)

[63] Franz, E., Jerichow, A., Möller, S., Pfitzmann, A., Stierand, I.: Computer based steganography: How it works and why therefore any restrictions on cryptography are nonsense, at best. In: Anderson, R.J. (ed.) Information Hiding (1st International Workshop), LNCS 1174, pp. 7–21. Springer-Verlag, Berlin Heidelberg (1996)

[64] Franz, E., Pfitzmann, A.: Einführung in die Steganographie und Ableitung eines neuen Stegoparadigmas [Introduction to steganography and derivation of a new stego-paradigm]. Informatik Spektrum **21**(4), 183–193 (1998)

[65] Franz, E., Pfitzmann, A.: Steganography secure against cover–stego-attacks. In: Pfitzmann, A. (ed.) Information Hiding (3rd International Workshop), LNCS 1768, pp. 29–46. Springer-Verlag, Berlin Heidelberg (2000)

[66] Franz, E., Schneidewind, A.: Adaptive steganography based on dithering. In: Proc. of ACM Multimedia and Security Workshop (MM&SEC), pp. 56–62. ACM Press, New York (2004)

[67] Franz, E., Schneidewind, A.: Pre-processing for adding noise steganography. In: Barni, M., Herrera-Joancomartí, J., Katzenbeisser, S., Pérez-González, F. (eds.) Information Hiding (7th International Workshop), LNCS 3727, pp. 189–203. Springer-Verlag, Berlin Heidelberg (2005)

[68] Fridrich, J.: Feature-based steganalysis for JPEG images and its implications for future design of steganographic schemes. In: Fridrich, J. (ed.) Information Hiding (6th International Workshop), LNCS 3200, pp. 67–81. Springer-Verlag, Berlin Heidelberg (2004)

[69] Fridrich, J.: Minimizing the embedding impact in steganography. In: Proc. of ACM Multimedia and Security Workshop (MM&SEC), pp. 2–10. Geneva, Switzerland (2006)

[70] Fridrich, J.: Steganography in Digital Media. Principles, Algorithms, and Applications. Cambridge University Press (2010)

[71] Fridrich, J., Filler, T.: Practical methods for minimizing embedding impact in steganography. In: Delp, E.J., Wong, P.W., Dittmann, J., Memon, N.D. (eds.) Security, Steganography and Watermarking of Multimedia Contents IX (Proc. of SPIE), vol. 6505. San Jose, CA (2007)

[72] Fridrich, J., Goljan, M.: Digital image steganography using stochastic modulation. In: Delp, E.J., Wong, P.W. (eds.) Security, Steganography and Watermarking of Multimedia Contents V (Proc. of SPIE), vol. 5022, pp. 191–202. San Jose, CA (2003)

[73] Fridrich, J., Goljan, M.: On estimation of secret message length in LSB steganography in spatial domain. In: Delp, E.J., Wong, P.W. (eds.) Security, Steganography and Watermarking of Multimedia Contents VI (Proc. of SPIE), pp. 23–34. San Jose, CA (2004)

[74] Fridrich, J., Goljan, M., Du, R.: Detecting LSB steganography in color and grayscale images. IEEE Multimedia $8$(4), 22–28 (2001)

[75] Fridrich, J., Goljan, M., Du, R.: Steganalysis based on JPEG compatibility. In: Tescher A. G., Vasudev, B., Bove Jr. V. M. (eds.) Multimedia Systems and Applications IV (Proc. of SPIE), pp. 275–280. Denver, CO (2001)

[76] Fridrich, J., Goljan, M., Hogea, D.: Attacking the OutGuess. In: Proc. of ACM Multimedia and Security Workshop (MM&SEC) (2000)

[77] Fridrich, J., Goljan, M., Hogea, D.: New methodology for breaking steganographic techniques for JPEGs. In: Delp, E.J., Wong, P.W. (eds.)

Security, Steganography and Watermarking of Multimedia Contents V (Proc. of SPIE), vol. 5022. San Jose, CA (2003)

[78] Fridrich, J., Goljan, M., Hogea, D.: Steganography of JPEG images: Breaking the F5 algorithm. In: Petitcolas, F.A.P. (ed.) Information Hiding (5th International Workshop 2002), LNCS 2578, pp. 310–323. Springer-Verlag, Berlin Heidelberg (2003)

[79] Fridrich, J., Goljan, M., Hogea, D., Soukal, D.: Quantitative steganalysis of digital images: Estimating the secret message length. ACM Multimedia Systems Journal **9**, 288–302 (2003)

[80] Fridrich, J., Goljan, M., Lisoněk, P., Soukal, D.: Writing on wet paper. In: Delp, E.J., Wong, P.W. (eds.) Security, Steganography and Watermarking of Multimedia Contents VII (Proc. of SPIE), vol. 5681, pp. 328–340. San Jose, CA (2005)

[81] Fridrich, J., Goljan, M., Lisoněk, P., Soukal, D.: Writing on wet paper. IEEE Transactions on Signal Processing **53**(10), 3923–3935 (2005)

[82] Fridrich, J., Goljan, M., Soukal, D.: Higher-order statistical steganalysis of palette images. In: Delp, E.J., Wong, P.W. (eds.) Security, Steganography and Watermarking of Multimedia Contents V (Proc. of SPIE), vol. 5020, pp. 178–190. San Jose, CA (2003)

[83] Fridrich, J., Goljan, M., Soukal, D.: Perturbed quantization steganography with wet paper codes. In: Proc. of ACM Multimedia and Security Workshop (MM&SEC), pp. 4–15. ACM Press, New York (2004)

[84] Fridrich, J., Goljan, M., Soukal, D.: Searching for the stego-key. In: Delp, E.J., Wong, P.W. (eds.) Security, Steganography and Watermarking of Multimedia Contents VI (Proc. of SPIE), vol. 5306, pp. 70–82. San Jose, CA (2004)

[85] Fridrich, J., Goljan, M., Soukal, D.: Efficient wet paper codes. In: Barni, M., Herrera-Joancomartí, J., Katzenbeisser, S., Pérez-González, F. (eds.) Information Hiding (7th International Workshop), LNCS 3727, pp. 204–218. Springer-Verlag, Berlin Heidelberg (2005)

[86] Fridrich, J., Kodowsky, J.: F5 has the 'best' embedding operation. Unpublished rump session talk at Information Hiding, Saint Malo, France (2007)

[87] Fridrich, J., Pevný, T., Kodovsky, J.: Statistically undetectable JPEG steganography: Dead ends, challenges, and opportunities. In: Proc. of ACM Multimedia and Security Workshop (MM&SEC), pp. 3–13. Dallas, Texas, USA (2007)

[88] Fridrich, J., Soukal, D.: Matrix embedding for large payloads. IEEE Transactions on Information Forensics and Security **1**(3), 390–394 (2006)

[89] Gloe, T., Kirchner, M., Winkler, A., Böhme, R.: Can we trust digital image forensics? In: Proc. of ACM Multimedia, pp. 78–86. ACM Press, New York, NY, USA (2007)

[90] Goljan, M., Fridrich, J., Du, R.: Distortion-free data embedding for images. In: Information Hiding, LNCS 2137, pp. 27–41. Springer-Verlag, Berlin Heidelberg (2001)

[91] Goljan, M., Fridrich, J., Holotyak, T.: New blind steganalysis and its implications. In: Delp, E.J., Wong, P.W. (eds.) Security, Steganography and Watermarking of Multimedia Contents VIII (Proc. of SPIE), vol. 6072. San Jose, CA (2006)

[92] Gross-Amblard, D.: Query-preserving watermarking of relational databases and XML documents. In: Proc. of ACM Symposium on Principles of Database Systems, pp. 191–201. New York, NY, USA (2003)

[93] Grothoff, C., Grothoff, K., Stutsman, R., Alkhutova, L., Atallah, M.: Translation-based steganography. In: Barni, M., Herrera-Joancomartí, J., Katzenbeisser, S., Pérez-González, F. (eds.) Information Hiding (7th International Workshop), LNCS 3727, pp. 213–233. Springer-Verlag, Berlin Heidelberg (2005). An extended version of July 2007 is available at http://grothoff.org/christian/lit.pdf (last access: October 2009)

[94] Guillon, P., Furon, T., Duhamel, P.: Applied public-key steganography. In: Wong, P.W., Delp, E.J. (eds.) Security, Steganography and Watermarking of Multimedia Contents IV (Proc. of SPIE), 4675. San Jose, CA (2002)

[95] Günther, P., Schönfeld, D., Winkler, A.: Reduced embedding complexity using BP message passing for LDGM codes. In: Delp, E.J., Wong, P.W., Dittmann, J., Memon, N.D. (eds.) Security, Steganography and Watermarking of Multimedia Contents X (Proc. of SPIE), vol. 6819. San Jose, CA (2008)

[96] Hamming, R.: Error detecting and error correcting codes. Bell System Technical Journal **29**(2), 147–160 (1950)

[97] Harmsen, J.J., Pearlman, W.A.: Steganalysis of additive noise modelable information hiding. In: Delp, E.J., Wong, P.W. (eds.) Security, Steganography and Watermarking of Multimedia Contents V (Proc. of SPIE), vol. 5022, pp. 131–142. San Jose, CA (2003)

[98] Harris, R.: Arriving at an anti-forensics consensus: Examining how to define and control the anti-forensics problem. Digital Investigation **3S**, 44–49 (2006)

[99] Hastad, J., Impagliazzo, R., Levin, L.A., Luby, M.: A pseudo-random generator from any one-way function. SIAM Journal on Computing **28**(4), 1364–1396 (1999)

[100] van Hateren, J.H., van der Schaaf, A.: Independent component filters of natural images compared with simple cells in primary visual cortex. Proceedings of the Royal Society London, Series B **265**, 359–366 (1998)

[101] Herrera-Moro, D.R., Rodríguez-Colín, R., Feregrino-Uribe, C.: Adaptive steganography based on textures. In: Proc. of IEEE International Conference on Electronics, Communications, and Computers, pp. 34–39 (2007)

[102] Hetzl, S., Mutzel, P.: A graph-theoretic approach to steganography. In: Dittmann, J., Katzenbeisser, S., Uhl, A. (eds.) Proc. of IFIP Communications and Multimedia Security, LNCS 3677, pp. 119–128. Springer-Verlag, Berlin Heidelberg (2005)

[103] Hipp, M.: MPG123 – fast MP3 player for Linux and UNIX systems (2001). http://www.mpg123.de (last access: October 2009)

[104] Hopper, N.J.: Towards a theory of steganography. Ph.D. thesis, Carnegie Mellon University, Pittsburgh, PA (2004)

[105] Hopper, N.J., Langford, J., Ahn, L.v.: Provable secure steganography. In: Yung, M. (ed.) Proc. of CRYPTO, LNCS 2442, pp. 77–92. Springer-Verlag, Berlin Heidelberg (2002)

[106] Huber, P.J.: Robust Statistics. Wiley, New York (1981)

[107] Huffman, D.A.: A method for the construction of minimum redundancy codes. Proceedings of the Institute of Radio Engineers (IRE) **40**(9), 1098–1101 (1952)

[108] Hundt, C., Liskiewicz, M., Wölfel, U.: Provably secure steganography and the complexity of sampling. In: Madria et al., S.K. (ed.) Algorithms and Computation (Proc. of ISAAC), vol. LNCS 4317, pp. 754–763. Springer-Verlag, Berlin Heidelberg (2006)

[109] IEC: 958, Digital Audio Interface. International Standard (1990)

[110] Ihaka, R., Gentlemen, R.: R – A language for data analysis and graphics. Journal of Computational Graphics and Statistics **5**, 299–314 (1996)

[111] Independent JPEG Group: (1998). http://www.ijg.org/ (last access: October 2009)

[112] ISO/IEC: 10918-1, Information Technology. Digital compression and coding of continuous-tone still images. International Standard (1994)

[113] ISO/IEC: 13818-3, Information Technology. Generic coding of moving pictures and associated audio: Audio. International Standard (1994)

[114] Johnson, N.F., Jajodia, S.: Steganalysis of images created using current steganography software. In: Pfitzmann, A. (ed.) Information Hiding (3rd International Workshop), LNCS 1768, pp. 273–289. Springer-Verlag, Berlin Heidelberg (2000)

[115] Kahn, D.: The history of steganography. In: Anderson, R.J. (ed.) Information Hiding (1st International Workshop), LNCS 1174, pp. 1–5. Springer-Verlag, Berlin Heidelberg (1996)

[116] Karhunen, K.: Zur Spektraltheorie Stochastischer Prozesse [On spectral theory of stochastic processes]. Annales Academiæ Scientiarum Fennicæ **37** (1946)

[117] Katzenbeisser, S., Petitcolas, F.A.P.: Defining security in steganographic systems. In: Delp, E.J., Wong, P.W. (eds.) Security, Steganography and Watermarking of Multimedia Contents IV (Proc. of SPIE), 4675, pp. 50–56. San Jose, CA (2002)

[118] Ker, A.D.: Improved detection of LSB steganography in grayscale images. In: Fridrich, J. (ed.) Information Hiding (6th International

Workshop), LNCS 3200, pp. 97–115. Springer-Verlag, Berlin Heidelberg (2004)

[119] Ker, A.D.: Quantitative evaluation of Pairs and RS steganalysis. In: Delp, E.J., Wong, P.W. (eds.) Security, Steganography and Watermarking of Multimedia Contents VI (Proc. of SPIE), vol. 5306, pp. 23–34. San Jose, CA (2004)

[120] Ker, A.D.: A general framework for the structural steganalysis of LSB replacement. In: Barni, M., Herrera-Joancomartí, J., Katzenbeisser, S., Pérez-González, F. (eds.) Information Hiding (7th International Workshop), LNCS 3727, pp. 296–311. Springer-Verlag, Berlin Heidelberg (2005)

[121] Ker, A.D.: Resampling and the detection of LSB matching in colour bitmaps. In: Delp, E.J., Wong, P.W. (eds.) Security, Steganography and Watermarking of Multimedia Contents VII (Proc. of SPIE), vol. 5681, pp. 1–15. San Jose, CA (2005)

[122] Ker, A.D.: Fourth-order structural steganalysis and analysis of cover assumptions. In: Delp, E.J., Wong, P.W. (eds.) Security, Steganography and Watermarking of Multimedia Contents VIII (Proc. of SPIE), vol. 6072, pp. 25–38. San Jose, CA (2006)

[123] Ker, A.D.: Batch steganography and pooled steganalysis. In: Camenisch, J., Collberg, C., Johnson, N.F., Sallee, P. (eds.) Information Hiding 2006 (8th International Workshop 2006), LNCS 4437, pp. 265–281. Springer-Verlag, Berlin Heidelberg (2007)

[124] Ker, A.D.: A capacity result for batch steganography. IEEE Signal Processing Letters **14**(8), 525–528 (2007)

[125] Ker, A.D.: Derivation of error distribution in least squares steganalysis. IEEE Transactions on Information Forensics and Security **2**(2), 140–148 (2007)

[126] Ker, A.D.: A fusion of maximum likelihood and structural steganalysis. In: Furon, T., Cayre, F., Doërr, G., Bas, P. (eds.) Information Hiding, LNCS 4567, pp. 204–219. Springer-Verlag, Berlin Heidelberg (2007)

[127] Ker, A.D.: The ultimate steganalysis benchmark. In: Proc. of ACM Multimedia and Security Workshop (MM&SEC), pp. 141–147. Dallas, Texas, USA (2007)

[128] Ker, A.D.: A weighted stego image detector for sequential LSB replacement. In: Workshop on Data Hiding for Information and Multimedia Security (Proc. of IAS), pp. 29–31. Manchester, UK (2007)

[129] Ker, A.D.: Perturbation hiding and the batch steganography problem. In: Solanki, K., Sullivan, K., Madhow, U. (eds.) Information Hiding, LNCS 5284, pp. 45–59. Springer-Verlag, Berlin Heidelberg (2008)

[130] Ker, A.D.: Steganographic strategies for a square distortion function. In: Delp, E.J., Wong, P.W., Dittmann, J., Memon, N.D. (eds.) Security, Steganography and Watermarking of Multimedia Contents X (Proc. of SPIE), vol. 6819. San Jose, CA (2008)

[131] Ker, A.D.: Estimating steganographic Fisher information in real images. In: Katzenbeisser, S., Sadeghi, A. (eds.) Information Hiding, LNCS 5806, pp. 73–88. Springer-Verlag, Berlin Heidelberg (2009)

[132] Ker, A.D.: Estimating the information theoretic optimal stego noise. In: Ho et al., A.T.S. (ed.) International Workshop on Digital Watermarking (IWDW 2009), LNCS 5703, pp. 184–198. Springer-Verlag, Berlin Heidelberg (2009)

[133] Ker, A.D., Böhme, R.: Revisiting weighted stego-image steganalysis. In: Delp, E.J., Wong, P.W., Dittmann, J., Memon, N.D. (eds.) Security, Forensics, Steganography and Watermarking of Multimedia Contents X (Proc. of SPIE), vol. 6819. San Jose, CA (2008)

[134] Ker, A.D., Pevný, T., Kodovský, J., Fridrich, J.: The square root law of steganographic capacity. In: Proc. of ACM Multimedia and Security Workshop (MM&SEC), pp. 107–116. Oxford, UK (2008)

[135] Kerckhoffs, A.: La cryptographie militaire. Journal des sciences militaires **IX**, 5–38, 161–191 (1883). http://www.petitcolas.net/fabien/kerckhoffs/crypto_militaire_1.pdf (last access: October 2009)

[136] Khanna, S., Zane, F.: Watermarking maps: Hiding information in structured data. In: SODA '00: Proc. of ACM–SIAM Symposium on Discrete Algorithms, pp. 596–605. Philadelphia, PA, USA (2000)

[137] Kharrazi, M., Sencar, H., Memon, N.: Cover selection for steganographic embedding. In: Proc. of IEEE ICIP, pp. 117–120 (2006)

[138] Kiayias, A., Raekow, Y., Russell, A.: Efficient steganography with provable security guarantees. In: Barni, M., Herrera-Joancomartí, J., Katzenbeisser, S., Pérez-González, F. (eds.) Information Hiding (7th International Workshop), LNCS 3727, pp. 118–130. Springer-Verlag, Berlin Heidelberg (2005)

[139] Kipper, G.: Investigator's Guide to Steganography. Auerbach, Boca Raton, FL (2003)

[140] Kirchner, M., Böhme, R.: Tamper hiding: Defeating image forensics. In: Furon, T., Cayre, F., Doërr, G., Bas, P. (eds.) Information Hiding, LNCS 4567, pp. 326–341. Springer-Verlag, Berlin Heidelberg (2007)

[141] Kirchner, M., Böhme, R.: Hiding traces of resampling in digital images. IEEE Transactions on Information Forensics and Security **3**, 582–592 (2008)

[142] Koops, B.J.: The Crypto Controversy. A Key Conflict in the Information Society. Kluwer, The Hague (1999)

[143] Koops, B.J.: Crypto law survey (2008). http://rechten.uvt.nl/koops/cryptolaw/index.htm (last access: October 2009)

[144] Köpsell, S.: Entwurf und Implementierung einer sicheren steganographischen Videokonferenz [Design and Implementation of a Secure Steganographic Video Conference]. Diploma thesis. Technische Universität Dresden, Institute for Theoretical Computer Science, Dresden, Germany (1999)

[145] Köpsell, S., Hillig, U.: How to achieve blocking resistance for existing systems enabling anonymous Web surfing. In: Atluri, V., Syverson, P.F., di Vimercati, S.D.C. (eds.) Proc. of ACM Workshop on Privacy in the Electronic Society (WPES), pp. 47–58. ACM Press, Washington, DC (2004)

[146] Kullback, S.: Information Theory and Statistics. Dover, New York (1968)

[147] Lam, E.Y.: Analysis of the DCT coefficient distributions for document coding. IEEE Signal Processing Letters $11(2)$, 97–100 (2004)

[148] Lam, E.Y., Goodman, J.W.: A mathematical analysis of the DCT coefficient distributions for images. IEEE Transactions on Image Processing $9(10)$, 1661–1666 (2000)

[149] Langley, P., Iba, W., Thompson, K.: An analysis of Bayesian classifiers. In: Proc. of 10th Conference on Artificial Intelligence, pp. 223–228. MIT Press (1992)

[150] Le, T.v., Kurosawa, K.: Efficient Public Key Steganography Secure Against Adaptively Chosen Stegotext Attacks. Report 2003/244. Cryptology ePrint Archive (2003). http://eprint.iacr.org/2003/244

[151] Lee, K., Jung, C., Lee, S., Lim, J.: New steganalysis methodology: LR cube analysis for the detection of LSB steganography. In: Barni, M., Herrera-Joancomartí, J., Katzenbeisser, S., Pérez-González, F. (eds.) Information Hiding (7th International Workshop), LNCS 3727, pp. 312–326. Springer-Verlag, Berlin Heidelberg (2005)

[152] Lee, K., Westfeld, A., Lee, S.: Category attack for LSB steganalysis of JPEG images. In: Shi Y Q., Jeon, B. (eds.) International Workshop on Digital Watermarking (IWDW 2006), LNCS 4283, pp. 35–48. Springer-Verlag, Berlin Heidelberg (2006)

[153] Lee, K., Westfeld, A., Lee, S.: Generalised category attack – improving histogram-based attack on JPEG LSB embedding. In: Furon, T., Cayre, F., Doërr, G., Bas, P. (eds.) Information Hiding, LNCS 4567, pp. 378–391. Springer-Verlag, Berlin Heidelberg (2007)

[154] Lie, W.N., Chang, L.C.: Data hiding in images with adaptive numbers of least significant bits based on the human visual system. In: Proc. of IEEE ICIP, vol. 1, pp. 286–290. Kobe, Japan (1999)

[155] Lin, E., Delp, E.: A review of fragile image watermarks. In: Proc. of ACM Multimedia and Security Workshop (MM&SEC), pp. 25–29. Orlando, FL (1999)

[156] Lindley, D.V.: Bayesian Statistics – A Review. Society for Industrial and Applied Mathematics (1995)

[157] Liu, Z., Ping, L., Chen, J., Wang, J., Pan, X.: Steganalysis based on differential statistics. In: Pointcheval, D., Mu, Y., Chen, K. (eds.) Cryptology and Network Security, LNCS 4301, pp. 224–240. Springer-Verlag, Berlin Heidelberg (2006)

[158] Loève, M.M.: Probability Theory. Van Nostrand, Princeton, NJ (1955)

[159] Lou, D.C., Sung, C.H.: A steganographic scheme for secure communications based on the chaos and Euler theorem. IEEE Transactions on Multimedia **6**, 501–509 (2004)

[160] Lu, P., Luo, X., Tang, Q., Shen, L.: An improved sample pairs method for detection of LSB embedding. In: Fridrich, J. (ed.) Information Hiding (6th International Workshop), LNCS 3200, pp. 116–127. Springer-Verlag, Berlin Heidelberg (2004)

[161] Lukáš, J., Fridrich, J.: Estimation of primary quantization matrix in double compressed JPEG images. In: Digital Forensic Research Workshop. Cleveland, OH (2003)

[162] Lukáš, J., Fridrich, J., Goljan, M.: Digital camera identification from sensor noise. IEEE Transactions on Information Forensics and Security **1**(2), 205–214 (2006)

[163] Lyu, S., Farid, H.: Detecting hidden messages using higher-order statistics and support vector machines. In: Petitcolas, F.A.P. (ed.) Information Hiding (5th International Workshop 2002), LNCS 2578, pp. 340–354. Springer-Verlag, Berlin Heidelberg (2003)

[164] Lyu, S., Farid, H.: Steganalysis using higher-order image statistics. IEEE Trans. on Information Forensics and Security **1**, 111–119 (2006)

[165] Malvar, K.S.: Extended lapped transforms: Properties, applications, and fast algorithms. IEEE Transactions on Signal Processing **40**(11), 2703–2714 (1992)

[166] Marvel, L.M., Boncelet, C.G.: Capacity of the additive steganographic channel. Mimeo (1999). Online available at http://www.eecis.udel.edu/~marvel/marvel_stegocap.ps.gz (last access: October 2009)

[167] Marvel, L.M., Boncelet, C.G., Retter, C.T.: Spread spectrum image steganography. IEEE Transactions on Image Processing **8**, 1075–1083 (1999)

[168] Marvel, L.M., Henz, B., Boncelet, C.G.: A performance study of ±1 steganalysis employing a realistic operating scenario. In: Proc. of Military Communications Conference (MILCOM). Orlando, FL (2007)

[169] Mielikainen, J.: LSB matching revisited. IEEE Signal Processing Letters **13**, 285–287 (2006)

[170] Moddemeijer, R.: On estimation of entropy and mutual information of continuous distributions. Signal Processing **16**(3), 233–246 (1989)

[171] Möller, B.: A public-key encryption scheme with pseudo-random ciphertexts. In: Samarati et al., P. (ed.) Computer Security (Proc. of ESORICS), LNCS 3193, pp. 335–351. Springer-Verlag, Berlin Heidelberg (2004)

[172] Moulin, P., O'Sullivan, J.A.: Information-theoretic analysis of information hiding. IEEE Transactions on Information Theory **49**(3), 563–593s (2003)

[173] Munuera, C.: Steganography and error-correcting codes. Signal Processing **87**, 1528–1533 (2007)

[174] Murdoch, S.J., Lewis, S.: Embedding covert channels in TCP/IP. In: Barni, M., Herrera-Joancomartí, J., Katzenbeisser, S., Pérez-González, F. (eds.) Information Hiding (7th International Workshop), LNCS 3727, pp. 247–261. Springer-Verlag, Berlin Heidelberg (2005)

[175] Natural Resources Conservation Service: NRCS photo gallery. `http://photogallery.nrcs.usda.gov` (last access: October 2009) (2008)

[176] Neyman, J., Pearson, E.: On the problem of the most efficient tests of statistical hypotheses. Philosophical Transactions of the Royal Society of London. Series A (Mathematical or Physical Character) **231**, 289–337 (1933)

[177] Ng, T.T., Chang, S.F., Lin, C.Y., Sun, Q.: Passive-blind image forensics. In: Zeng, W., H.Yu, Lin, C.Y. (eds.) Multimedia Security Technologies for Digital Rights, pp. 383–412. Academic Press (2006)

[178] Nilsson, M.: ID3v2 – The audience is informed (1998). `http://www.id3.org` (last access: October 2009)

[179] Özer, H., Avcibaş, I., Sankur, B., Memon, N.D.: Steganalysis of audio based on audio quality metrics. In: Delp, E.J., Wong, P.W. (eds.) Security, Steganography and Watermarking of Multimedia Contents III (Proc. of SPIE), vol. 5020, pp. 55–66. San Jose, CA (2003)

[180] Park, Y.R., Kang, H.H., Shin, S.U., Kown, K.R.: A steganographic scheme in digital images using information of neighboring pixels. In: Wang, L., Chen, K., Ong, Y.S. (eds.) Advances in Natural Computation (Proc. of ICNC), LNCS 3612, pp. 962–967. Springer-Verlag, Berlin Heidelberg (2005)

[181] Parnas, D.L.: The secret history of information hiding. In: Software Pioneers: Contributions to Software Engineering, pp. 399–409. Springer-Verlag, Berlin Heidelberg (2002)

[182] Pearl, J.: Probabilistic Reasoning in Intelligent Systems. Morgan Kaufmann (1988)

[183] Pennebaker, W.B., Mitchell, J.L.: JPEG: Still Image Data Compression Standard. 6th edn. Kluwer, Boston (2001)

[184] Petitcolas, F.A.P.: MP3Stego (2002). `http://www.cl.cam.ac.uk/~fapp2/steganography/mp3stego/` (last access: October 2009)

[185] Petitcolas, F.A.P., Anderson, R.J., Kuhn, M.G.: Information hiding – a survey. Proceedings of the IEEE, Special issue on Protection of Multimedia Content **87**, 1062–1078 (1999)

[186] Pevný, T.: Kernel methods in steganalysis. Ph.D. thesis, Graduate School of Binghamton University, State University of New York (2008)

[187] Pevný, T., Fridrich, J.: Merging Markov and DCT features for multiclass JPEG steganalysis. In: Delp, E.J., Wong, P.W., Dittmann, J., Memon, N.D. (eds.) Security, Steganography and Watermarking of Multimedia Contents IX (Proc. of SPIE), vol. 6505. San Jose, CA (2007)

[188] Pevný, T., Fridrich, J.: Benchmarking for steganography. In: Solanki, K., Sullivan, K., Madhow, U. (eds.) Information Hiding, LNCS 5284, pp. 251–267. Springer-Verlag, Berlin Heidelberg (2008)

[189] Pevný, T., Fridrich, J.: Detection of double-compression in JPEG images for applications in steganography. IEEE Transactions on Information Forensics and Security **3**(2), 247–258 (2008)

[190] Pevný, T., Fridrich, J.: Multiclass detector of current steganographic methods for JPEG format. IEEE Transactions on Information Forensics and Security **3**, 635–650 (2008)

[191] Pevný, T., Fridrich, J.: Novelty detection in blind steganalysis. In: Proc. of ACM Multimedia and Security Workshop (MM&SEC), pp. 167–176. Oxford, UK (2008)

[192] Pfitzmann, A., Hansen, M.: Anonymity, unlinkability, undetectability, unobservability, pseudonymity, and identity management – a consolidated proposal for terminology. `http://dud.inf.tu-dresden.de/Anon_Terminology.shtml` (2008). (Version 0.31e)

[193] Pfitzmann, B.: Information hiding terminology. In: Anderson, R.J. (ed.) Information Hiding (1st International Workshop), LNCS 1174, pp. 347–350. Springer-Verlag, Berlin Heidelberg (1996)

[194] Plackett, R.L.: Karl Pearson and the chi-squared test. International Statistical Review **51**(1), 59–72 (1983)

[195] Platt, C.: UnderMP3Cover (2004). `http://sourceforge.net/projects/ump3c` (last access: October 2009)

[196] Plonka, G., Tasche, M.: Invertible integer DCT algorithms. Applied and Computational Harmonic Analysis **15**(1), 70–88 (2003)

[197] Popper, K.R.: Logik der Forschung [The Logic of Scientific Discovery]. Springer-Verlag, Wien (1935). Translation to English in 1959

[198] Provos, N.: Defending against statistical steganalysis. In: Proc. of 10th USENIX Security Symposium. Washington, DC (2001)

[199] R Development Core Team: R: A language and environment for statistical computing. R Foundation for Statistical Computing, Vienna, Austria (2005). `http://www.R-project.org`

[200] Ramanath, R., Snyder, W.E., Bilbro, G.L., Sander III, W.A.: Demosaicking methods for Bayer color arrays. Journal of Electronic Imaging **11**(3), 306–315 (2002)

[201] Ripley, B.D.: Pattern Recognition and Neural Networks. Cambridge University Press (1996)

[202] Roue, B., Bas, P., Chassery, J.M.: Improving LSB steganalysis using marginal and joint probabilistic distributions. In: Proc. of ACM Multimedia and Security Workshop (MM&SEC), pp. 75–80. ACM Press, New York (2004)

[203] Royston, P.: An extension of Shapiro and Wilk's test for normality to large samples. Applied Statistics **31**, 115–124 (1982)

[204] Ru, X.M., Zhang, H.J., Huang, X.: Steganalysis of audio: Attacking the StegHide. In: Proc. International Conference on Machine Learning and Cybernetics, pp. 3937–3942. Guangzhou, China (2005)

[205] Salamon, C., Corney, J.: Information hiding through variance of the parametric orientation underlying a B-rep face. In: Solanki, K., Sullivan, K., Madhow, U. (eds.) Information Hiding, LNCS 5284, pp. 268–282. Springer-Verlag, Berlin Heidelberg (2008)

[206] Sallee, P.: Model-based steganography. In: Kalker et al., T. (ed.) International Workshop on Digital Watermarking (IWDW 2003), LNCS 2939, pp. 154–167. Springer-Verlag, Berlin Heidelberg (2004)

[207] Sallee, P.: Model-based methods for steganography and steganalysis. International Journal of Image an Graphics **5**(1), 167–189 (2005)

[208] Salomon, D.: Data Compression: The Complete Reference. Second edn. Springer-Verlag, Berlin Heidelberg (2000). Extra material on decorrelation, online available at http://www.davidsalomon.name/ DC2advertis/DeCorr.pdf (last access: October 2009)

[209] Schneier, B.: Beyond Fear: Thinking Sensibly about Security in an Uncertain World. Copernicus Books (Springer-Verlag) (2003)

[210] Schönfeld, D.: Einbetten mit minimaler Werkänderung [Embedding with minimal cover changes]. Datenschutz und Datensicherheit **25**(11), 666–671 (2001)

[211] Schönfeld, D., Winkler, A.: Embedding with syndrome coding based on BCH codes. In: Proc. of ACM Multimedia and Security Workshop (MM&SEC), pp. 214–223. Geneva, Switzerland (2006)

[212] Schönfeld, D., Winkler, A.: Reducing the complexity of syndrome coding for embedding. In: Furon, T., Cayre, F., Doërr, G., Bas, P. (eds.) Information Hiding, LNCS 4567, pp. 145–158. Springer-Verlag, Berlin Heidelberg (2007)

[213] Shannon, C.: Communication theory of secrecy systems. Bell System Technical Journal **28**, 656–715 (1949)

[214] Sharp, T.: An implementation of key-based digital signal steganography. In: Moskowitz, I.S. (ed.) Information Hiding (3rd International Workshop), LNCS 2137, pp. 13–26. Springer-Verlag, Berlin Heidelberg (2001)

[215] Shi, Y.Q., Chen, C., Chen, W.: A Markov process based approach to effective attacking JPEG steganography. In: Camenisch, J., Collberg, C., Johnson, N.F., Sallee, P. (eds.) Information Hiding 2006 (8th International Workshop 2006), LNCS 4437, pp. 249–264. Springer-Verlag, Berlin Heidelberg (2007)

[216] Shikata, J., Matsumoto, T.: Unconditionally secure steganography against active attacks. IEEE Transactions on Information Theory **54**(6), 2690–2705 (2008)

[217] Simmons, G.J.: The prisoners' problem and the subliminal channel. In: Chaum, D. (ed.) Proceedings of CRYPTO, pp. 51–67. Santa Barbara, CA (1983)

[218] Solanki, K., Sarkar, A., Manjunath, B.S.: YASS: Yet another stegano-graphic scheme that resists blind steganalysis. In: Furon, T., Cayre, F., Doërr, G., Bas, P. (eds.) Information Hiding, LNCS 4567, pp. 16–31. Springer-Verlag, Berlin Heidelberg (2007)

[219] Solanki, K., Sullivan, K., Madhow, U., Manjunath, B., Chandrasekaran, S.: Statistical restoration for robust and secure steganography. In: Proc. of IEEE ICIP, vol. 2, pp. 1118–1121 (2005)

[220] Solanki, K., Sullivan, K., Madhow, U., Manjunath, B., Chandrasekaran, S.: Provably secure steganography: Achieving zero K-L divergence using statistical restoration. In: Proc. of IEEE ICIP, pp. 125–128 (2006)

[221] Srivastava, A., Lee, A.B., Simoncelli, E.P., Zhu, S.C.: On advances in statistical modeling of natural images. Journal of Mathematical Imaging and Vision **18**, 17–33 (2003)

[222] Stego-Lame: (2002). `http://sourceforge.net/projects/stego-lame` (page unavailable in May 2008)

[223] Tanabe, K., Sagae, M.: An exact Cholesky decomposition and the generalized inverse of the variance-covariance matrix of the multinomial distribution, with applications. Journal of the Royal Statistical Society, Series B (Methodological) **54**(1), 211–219 (1992)

[224] Taylor, J., Verbyla, A.: Joint modelling of location and scale parameters of the $t$ distribution. Statistical Modelling **4**, 91–112 (2004)

[225] Thévenaz, P., Blu, T., Unser, M.: Image interpolation and resampling. In: Bankman, I.N. (ed.) Handbook of Medical Imaging, Processing and Analysis, chap. 25, pp. 393–420. Academic Press, San Diego, CA, USA (2000)

[226] Topkara, M., Topkara, U., Atallah, M.J., Taskiran, C., Lin, E., Delp, E.J.: A hierarchical protocol for increasing the stealthiness of stegano-graphic methods. In: Proc. of ACM Multimedia and Security Workshop (MM&SEC), pp. 16–24. ACM Press, New York (2004)

[227] Ullerich, C.: Analyse modellbasierter Steganographie [Analysis of Model-based Steganography]. Diploma thesis. Technische Universität Dresden, Institute of Systems Architecture, Dresden, Germany (2007). Published in 2008 by VDM Verlag, Saarbrücken

[228] Ullerich, C., Westfeld, A.: Weaknesses of MB2. In: Shi, Y.Q., Kim, H.J., Katzenbeisser, S. (eds.) Digital Watermarking (Proc. of IWDW 2007), LNCS 5041, pp. 127–142. Springer-Verlag, Berlin Heidelberg (2008)

[229] Venables, W.N., Ripley, B.D.: Modern Applied Statistics with S. Fourth edn. Springer-Verlag, Berlin Heidelberg (2002)

[230] Wang, Y., Moulin, P.: Steganalysis of block-DCT image steganography. In: Proc. of IEEE Workshop on Statistical Signal Processing, pp. 339–342 (2003)

[231] Wang, Y., Moulin, P.: Perfectly secure steganography: Capacity, error exponents, and code constructions. IEEE Transactions on Information Theory **54**(6), 2706–2722 (2008)

[232] Wayner, P.: Mimic functions. Cryptologia **16**, 193–214 (1992)

[233] Westfeld, A.: F5 – A steganographic algorithm. In: Information Hiding (3rd International Workshop), LNCS 2137, pp. 289–302. Springer-Verlag, Berlin Heidelberg (2001)

[234] Westfeld, A.: Detecting low embedding rates. In: Petitcolas, F.A.P. (ed.) Information Hiding (5th International Workshop 2002), LNCS 2578, pp. 324–339. Springer-Verlag, Berlin Heidelberg (2003)

[235] Westfeld, A.: ROC curves for steganalysts. Mimeo (2007). Online available at http://os.inf.tu-dresden.de/~westfeld/publikationen/westfeld.wacha07.pdf (last access: October 2009)

[236] Westfeld, A.: Steganography for radio amateurs. A DSSS-based approach for slow scan television. In: Information Hiding (8th International Workshop 2006), LNCS 4437, pp. 201–215. Springer-Verlag, Berlin Heidelberg (2007)

[237] Westfeld, A.: Generic adoption of spatial steganalysis to transformed domain. In: Solanki, K., Sullivan, K., Madhow, U. (eds.) Information Hiding, LNCS 5284, pp. 161–177. Springer-Verlag, Berlin Heidelberg (2008)

[238] Westfeld, A., Pfitzmann, A.: Attacks on steganographic systems. In: Pfitzmann, A. (ed.) Information Hiding (3rd International Workshop), LNCS 1768, pp. 61–76. Springer-Verlag, Berlin Heidelberg (2000)

[239] Willems, F.M.J., van Dijk, M.: Capacity and codes for embedding information in gray-scale signals. IEEE Transactions on Information Theory $51$(3), 1209–1214 (2005)

[240] Witten, I.H., Neal R., M., Cleary, J.G.: Arithmetic coding for data compression. Communications of the ACM $20$, 520–540 (1987)

[241] Wu, D.C., Tsai, W.H.: A steganographic method for images by pixel-value differencing. Pattern Recognition Letters $24$, 1613–1626 (2003)

[242] Wu, D.C., Wu, N.I., Tsai, C.S., Hwang, M.S.: Image steganographic scheme based on pixel-value differencing and LSB replacement methods. IEE Proc. on Vision, Image and Signal Processing $152$(5), 611–615 (2005)

[243] Wu, M., Zhu, Z., Jin, S.: Detection of hiding in the LSB of DCT coefficients. In: Huang, D.S., Zhang, X.P., Huang, G.B. (eds.) Proc. of Advances in Intelligent Computing (ICIC), LNCS 3644, pp. 291–300. Springer-Verlag, Berlin Heidelberg (2005)

[244] Wu, M., Zhu, Z., Jin, S.: A new steganalytic algorithm for detecting JSteg. In: Lu, X., Zhao, W. (eds.) Proc. of Networking and Mobile Computing (ICCNMC), LNCS 3619, pp. 1073–1082. Springer-Verlag, Berlin Heidelberg (2005)

[245] Yang, C.H., Weng, C.Y., Wang, S.J., Sun, H.M.: Adaptive data hiding in edge areas of images with spatial LSB domain systems. IEEE Transactions on Information Forensics and Security $3$(3), 488–497 (2008)

[246] Young, A., Yung, M.: Kleptography: Using cryptography against cryptography. In: Funny, W. (ed.) Advances in Cryptology. Proc. of

EUROCRYPT, LNCS 1233, pp. 62–74. Springer-Verlag, Berlin Heidelberg (1997)

[247] Yu, X., Tan, T., Wang, Y.: Extended optimization method of LSB steganalysis. In: Proc. IEEE International Conference on Image Processing, vol. 2, pp. 1102–1105 (2005)

[248] Yu, X., Wang, Y., Tan, T.: On estimation of secret message length in JSteg-like steganography. In: Proc. of International Conference on Pattern Recognition (ICPR), vol. 4, pp. 673–676. IEEE Press (2004)

[249] Yu, X., Zhu, X., Babaguchi, N.: Steganography using sensor noise and linear prediction synthesis filter. In: Proc. of IEEE ICIP, vol. 2, pp. 157–160 (2007)

[250] Zhang, R., Sachnev, V., Kim, H.J.: Fast BCH syndrome coding for steganography. In: Katzenbeisser, S., Sadeghi, A. (eds.) Information Hiding, LNCS 5806, pp. 48–58. Springer-Verlag, Berlin Heidelberg (2009)

[251] Zhang, T., Ping, X.: A fast and effective steganalytic technique against JSteg-like algorithms. In: Proc. of ACM Symposium on Applied Computing, pp. 307–311. ACM Press, Melbourne, Florida (2003)

[252] Zhang, T., Ping, X.: A new approach to reliable detection of LSB steganography in natural images. Signal Processing **83**, 2085–2093 (2003)

[253] Zhang, W., Li, S.: Security measurements of steganographic systems. In: Jakobsson, M., Yung, M., Zhou, J. (eds.) Applied Cryptography and Network Security (ACNS), LNCS 3089, pp. 194–204. Springer-Verlag, Berlin Heidelberg (2004)

[254] Zhang, W., Wang, S., Zhang, X.: Improving embedding efficiency of covering codes for applications in steganography. IEEE Communications Letters **11**(8), 680–682 (2007)

[255] Zhang, W., Zhang, X., Wang, S.: A double layered "plus-minus one" data embedding scheme. IEEE Signal Processing Letters **14**(11), 848–851 (2007)

[256] Zhang, X., Wang, S.: Vulnerability of pixel-value differencing steganography to histogram analysis and modification for enhanced security. Pattern Recognition Letters **25**, 331–339 (2004)

[257] Zhang, X., Wang, S., Zhang, K.: Steganography with least histogram abnormality. In: Gorodetsky, V., Popyack, L.J., Skormin, V.A. (eds.) Computer Network Security (MMM-ACNS), LNCS 2776, pp. 395–406. Springer-Verlag, Berlin Heidelberg (2003)

[258] Zöllner, J., Federrath, H., Klimant, H., Pfitzmann, A., Piotraschke, R., Westfeld, A., Wicke, G., Wolf, G.: Modeling the security of steganographic systems. In: Aucsmith, D. (ed.) Information Hiding (2nd International Workshop), LNCS 1525, pp. 306–318. Springer-Verlag, Berlin Heidelberg (1998)

# List of Tables

# List of Figures

# List of Acronyms

AC ........ alternating current
AUC ...... area under the curve
BBN ...... Bayesian belief network
BCH ...... Bose–Ray-Chaudhuri–Hocquenghem (cyclic error-correction codes)
CAD ...... computer aided design
CAS ....... chaotic asymmetric steganography
CBR ...... constant bit rate
CD ........ compact disc
CDF ...... cumulative density function
CFA ....... colour filter array
CLT ....... central limit theorem
COM ...... centre of mass
CRC ...... cyclic redundancy check
DC ........ direct current
DCT ...... discrete cosine transformation
DFIL ...... Digital Forensic Image Library (an image provider)
DFT ...... discrete Fourier transformation
DHIES .... Diffie–Hellman integer encryption scheme
DWT ...... discrete wavelet transformation
EER ....... equal error rate
EM ........ expectation maximisation
FDCT ..... fast discrete cosine transformation
FFT ....... fast Fourier transformation
FLD ....... Fisher linear discriminant analysis
HCF ...... histogram characteristic function
IDCT ..... inverse discrete cosine transformation
IDS ....... intrusion detection system
IID ........ independent and identically-distributed
IQR ....... interquartile range
ISO ....... International Organization for Standardization

IWLS ..... iterative reweighted least squares
JPEG ..... Joint Photographic Experts Group
KLD ...... Kullback–Leibler divergence
LAHD ..... local angular harmonic decomposition
LDGM .... low-density generator matrix
LL ........ log likelihood
LRT ....... likelihood ratio test
LSB ....... least significant bit
LSM ....... least-squares method
MAC ...... message authentication code
MAE ...... mean absolute error
MB1 ...... model-based steganography I
MB2 ...... model-based steganography II (with blockiness correction)
MDCT .... modulated discrete cosine transformation
MDS ...... multi-dimensional scaling
ML ........ maximum likelihood
MLT ...... modulated lapped transformation
MP ........ matching pursuit
MP3 ....... ISO/MPEG 1 Audio Layer-3
MPEG .... Moving Picture Expert Group
MSE ...... mean square error
NBC ...... naïve Bayes classifier
NP ........ nondeterministic polynomial-time
NRCS .... National Resources Conservation Service (an image provider)
OLS ....... ordinary least-squares
PBR ...... probabilistic bias removal
PCA ...... principal component analysis
PCM ..... pulse code modulation
PDF ....... probability density function
PKS ....... public-key steganography
PM1 ....... plus-minus-one (embedding)
PQ ........ perturbed quantisation
PRNG ..... pseudorandom number generator
PSNR ..... peak signal to noise ratio
PVD ...... pixel-value differencing
QDA ...... quadratic discriminant analysis
QIM ....... quantisation index modulation
RGB ...... red, green, blue
ROC ...... receiver operating characteristics
RS ........ regular/singular steganalysis
RSA ....... Rivest–Shamir–Adleman (asymmetric crypto system)
SCFSI ..... scale factor selection information
SCMS ..... serial copy management system
SKS ....... secret-key steganography
SPA ....... sample pair analysis

SQAM .... sound quality assessment material
SSIS ....... spread spectrum image steganography
SVM ...... support vector machine
TMDE .... total minimal decision error
VBR ...... variable bit rate
WLS ...... weighted least squares
WPC ...... wet paper codes
WS ........ weighted stego image (steganalysis)
XML ...... extensible mark-up language

# List of Symbols

| | |
|---|---|
| $\mathbf{1}$ | matrix of ones |
| $\alpha$ | false positive probability |
| $\beta$ | missing probability $(1-$probability of detection$)$ |
| $\chi^2$ | chi-squared statistic of Pearson's contingency test |
| $\chi^2_{\mathrm{crit}}$ | critical value for chi-squared of Pearson's contingency test |
| $\delta$ | Kronecker delta |
| $\Delta$ | adj. difference between cardinality of regular and singular groups in RS |
| $\varepsilon$ | bound for $\varepsilon$-security |
| $\epsilon_i$ | error of $i$th trace set in SPA/LSM, prediction error |
| $\epsilon$ | residual in regression analysis |
| $\eta$ | embedding efficiency (general) |
| $\eta_{\#}$ | embedding efficiency (by number of changes) |
| $\eta_{\odot}$ | embedding efficiency (by wet pixel constraint) |
| $\Gamma$ | gamma function |
| $\iota$ | number of iterations per cover, secondary feature index |
| $\kappa_i$ | criterion to calculate $i$th symbolic feature |
| $\lambda$ | parameter of weighted stego image, parameter of generalised Cauchy distribution |
| $\Lambda$ | likelihood ratio test score |
| $\mu$ | mean, location parameter |
| $\nu$ | degrees of freedom parameter of Student $t$ distribution |
| $\omega_{\mathrm{eff}}$ | effective bit rate |
| $\omega_{\mathrm{nom}}$ | nominal bit rate |
| $\Omega$ | sample space of probability space of cover generating process |
| $\dot{\Omega}$ | set of all uncompressed (original) covers |
| $\boldsymbol{\Phi}$ | $n \times n$ matrix indicating pixel neighbourhood |
| $\pi$ | radial constant, parameter of binomial distribution, probabilities for indirect cover models |

| | |
|---|---|
| $\sigma$ | standard deviation, scale parameter, parameter of generalised Cauchy distribution |
| $\boldsymbol{\Sigma}$ | covariance or correlation matrix |
| $\varrho$ | correlation coefficient of constant correlation image model |
| $\tau$ | detection threshold |
| $\boldsymbol{\theta}$ | parameter set of classifier |
| $\vartheta$ | normalisation factor for weights in WS analysis |
| $\boldsymbol{a}$ | transformation matrix, mixing matrix in constant correlation image model |
| $\boldsymbol{a}_{1\mathrm{D}}$ | 1D-DCT transformation matrix |
| $\boldsymbol{a}_{2\mathrm{D}}$ | 2D-DCT transformation matrix |
| $\hat{\boldsymbol{a}}$ | vector of estimated regression coefficients (location model) |
| $\boldsymbol{a}_{\diamond}$ | sub-matrix of mixing matrix |
| $b_i$ | regression weights |
| $b_u^{i,j}$ | $u$th low-precision histogram bin of DCT subband $(i,j)$ |
| $\breve{b}_i$ | type of $i$th block in MP3 file |
| $\boldsymbol{b}_i$ | $i$th block in MP3 file |
| $\hat{\boldsymbol{b}}$ | vector of estimated regression coefficients |
| $\boldsymbol{b}^{(0)}$ | steganographic semantic of cover block |
| $\boldsymbol{b}^{(1)}$ | steganographic semantic of stego block |
| $\mathcal{B}$ | binomial distribution, set of all MP3 blocks |
| $c$ | class symbol (output of classifier) |
| $\boldsymbol{c}_R, \boldsymbol{c}_B$ | Cr and Cb components of YCrCb colour model |
| $\mathcal{C}_i$ | trace sets in SPA |
| $d_j$ | histogram bin of big values distribution over MP3 blocks |
| $\hat{d}$ | estimated bias in WS analysis |
| $\boldsymbol{d}$ | parity check matrix |
| $\overline{\boldsymbol{d}}$ | reduced parity check matrix (columns of wet samples removed) |
| $\mathcal{D}_i$ | pre-cover trace sets in SPA/ML |
| $e$ | energy |
| $\mathsf{E}$ | expected value operator |
| $\mathcal{E}$ | trace subset of even pairs in SPA |
| $\boldsymbol{f}$ | feature vector |
| $\mathcal{F}$ | feature domain |
| $\boldsymbol{g}_j^{(i)}$ | $j$th granule of $i$th frame |
| $h$ | realisation of unobservable random variable in mixture models |
| $\hat{h}$ | realisation of unobservable random variable estimated from the stego object |
| $h_v^{i,j}$ | $v$th high-precision histogram bin of DCT subband $(i,j)$ |
| $\tilde{h}^{i,j}$ | expected frequency for high-precision histogram bin of DCT subband $(i,j)$ |
| $H$ | unobservable random variable in mixture models |

| | |
|---|---|
| $i$ | sample index, cover index, encoder index, frame index, subband index |
| $\boldsymbol{I}$ | identity matrix |
| $j$ | message bit index, message index, layer index, feature index, subband index |
| $k$ | number of blocks, parameter in mod-$k$ embedding, group size in RS analysis, order of structural steganalysis, number of layers $\mathcal{L}$, dimension of projection space for indirect cover models, number of features |
| $\boldsymbol{k}$ | secret key |
| $\boldsymbol{k}_{\mathrm{priv}}$ | private key |
| $\boldsymbol{k}_{\mathrm{pub}}$ | public key |
| $\mathcal{K}$ | key space |
| $l$ | number of message bits per block in syndrome coding |
| $\ell$ | security parameter |
| $\mathcal{L}$ | layer of similar predictability |
| $m$ | message bit |
| $\boldsymbol{m}$ | secret message |
| $\mathfrak{m}$ | mask vector in RS analysis |
| $\mathcal{M}$ | message space |
| $n$ | cover size in samples |
| $n_{\square}$ | block size of linear block code |
| $\mathfrak{n}$ | large semi-prime |
| $N$ | number of covers, block size in pixels, number of subsets, number of encoders |
| $\mathcal{N}$ | univariate Gaussian distribution |
| $\boldsymbol{\mathcal{N}}$ | multivariate Gaussian distribution |
| $\mathbb{N}$ | set of all natural numbers |
| $\boldsymbol{o}$ | distance matrix for multi-dimensional scaling |
| $\mathcal{O}$ | Landau symbol, trace subset of odd pairs in SPA |
| $p$ | net embedding rate |
| $\hat{p}$ | estimated net embedding rate |
| $\hat{p}_i$ | embedding rate estimate of $i$th trace set in SPA |
| $\mathfrak{p}$ | large prime |
| $\hat{P}$ | estimation error in quantitative steganalysis |
| $\mathcal{P}_0$ | probability distribution of cover objects |
| $\mathcal{P}_1$ | probability distribution of stego objects |
| $\mathfrak{P}$ | power set operator |
| $q$ | JPEG compression quality, proportion of correctly classified feature objects |
| $\hat{q}$ | estimated JPEG compression quality from spatial domain stego object |
| $\boldsymbol{q}$ | diagonal matrix of quantisation factors |
| $\overline{\boldsymbol{q}}$ | diagonal matrix of inverse quantisation factors |
| $\mathcal{Q}_{\mathrm{SCFSI}}$ | set of possible combinations of SCFSI flags in MP3 |

| | |
|---|---|
| $\mathbb{Q}$ | set of all rational numbers |
| $\boldsymbol{R}$ | random vector in embedding operation, independent random variables in constant correlation image model |
| $\mathcal{R}_{\mathfrak{m}}$ | multi-set of regular groups under mask $\mathfrak{m}$ in RS analysis |
| $\mathbb{R}$ | set of all real numbers |
| $s$ | ciphertext |
| $s_{\mathrm{eff},i}^{\mathrm{fr}}$ | effective size of $i$th MP3 frame |
| $\mathcal{S}$ | infinite set of real and imaginary natural phenomena, test set for classifier |
| $\mathcal{S}_{\mathfrak{m}}$ | multi-set of singular groups under mask $\mathfrak{m}$ in RS analysis |
| $\boldsymbol{t}_k$ | $k$th order transition matrix in higher-order structural steganalysis |
| $\mathcal{T}$ | training set for classifier |
| $u$ | index of low-precision histogram bin, weight parameter in WS analysis |
| $\boldsymbol{u}$ | matrix of explanatory variables (location model) |
| $\mathcal{U}$ | continuous uniform distribution |
| $\ddot{\mathcal{U}}$ | discretised uniform distribution |
| $v$ | index of high-precision histogram bin |
| $\boldsymbol{v}$ | difference vector in syndrome coding, matrix of explanatory variables, sub-vector of mixing matrix |
| $\overline{\boldsymbol{v}}$ | reduced difference vector in syndrome coding |
| $\tilde{w}_j$ | average weight of pixels in $j$th layer $\mathcal{L}_j$ |
| $\boldsymbol{w}$ | weights in WS analysis |
| $x$ | sample |
| $\overline{x}$ | sample with flipped LSB |
| $x^{(0)}$ | cover sample |
| $x^{(1)}$ | sample carrying steganographic semantic |
| $\boldsymbol{x}$ | signal |
| $\boldsymbol{x}_{i(,j)}$ | signal encoded with $\mathsf{Encode}_i$ (double-compressed with sequence $(i,j)$) |
| $\overline{\boldsymbol{x}}$ | signal with flipped LSB |
| $\dot{\boldsymbol{x}}$ | uncompressed (original) signal |
| $\boldsymbol{x}_{\boxplus}$ | blockwise signal |
| $\boldsymbol{x}^{(0)}$ | cover |
| $\boldsymbol{x}^{(1)}$ | stego object |
| $\boldsymbol{x}^{(\boldsymbol{m})}$ | stego object carrying message $\boldsymbol{m}$ |
| $\boldsymbol{x}^{(p)}$ | stego object with net embedding rate $p$ |
| $\boldsymbol{X}_{\mathrm{det}}$ | deterministic part of cover |
| $\boldsymbol{X}_{\mathrm{indet}}$ | indeterministic part of cover |
| $\mathcal{X}$ | cover symbol alphabet |
| $\mathcal{X}^*$ | set of all covers |
| $\mathcal{X}^n$ | set of all covers of size $n$ |
| $\boldsymbol{y}$ | signal in transformed domain, response in regression analysis |
| $\boldsymbol{y}_{\boxplus}$ | blockwise signal in transformed domain |

| | |
|---|---|
| $\boldsymbol{y}_{\boxplus}^{*}$ | blockwise signal in transformed domain, quantised |
| $\tilde{\boldsymbol{y}}_{\boxplus}$ | blockwise joint error due to rounding and stego noise |
| $Y$ | random variable in mixture models |
| $z$ | projection dimensions in example cover model ('world view') |
| $Z$ | random variable in mixture models |
| $Z_{\text{cover}}$ | between-image estimation error |
| $Z_{\text{flips}}$ | estimation error due to number of changes (component of $Z_{\text{message}}$) |
| $Z_{\text{message}}$ | within-image estimation error |
| $Z_{\text{pos}}$ | estimation error due to position of changes (component of $Z_{\text{message}}$) |
| $\boldsymbol{Z}^{(0)}$ | random variable of projected covers |
| $\boldsymbol{Z}^{(1)}$ | random variable of projected stego objects |
| $\mathcal{Z}^{k}$ | set of cover projections after dimension reduction |
| $\mathbb{Z}$ | set of all integers |
| $\mathbb{Z}_{k}$ | Galois field modulo $k$ |
| $\#_{\text{bl}}$ | number of blocks in an MP3 file |
| $\#_{\text{fr}}$ | number of frames in an MP3 file |

# List of Functions

# Index